Practical Laboratory Automation

Practical Laboratory Automation

Made Easy with AutoIt

Matheus C. Carvalho

Verlag GmbH & Co. KGaA

Author

Dr. Matheus C. Carvalho
Southern Cross University
Centre for Coastal Biogeochemistry Res.
PO Box 157, Lismore, NSW, 2480
2480 Lismore
Australia

Cover
Microplate / fotolia.com © Caleb Foster;
Vials autosampler tray / fotolia.com © bigy9950

All books published by **Wiley-VCH** are carefully produced. Nevertheless, authors, editors, and publisher do not warrant the information contained in these books, including this book, to be free of errors. Readers are advised to keep in mind that statements, data, illustrations, procedural details or other items may inadvertently be inaccurate.

Library of Congress Card No.: applied for

British Library Cataloguing-in-Publication Data
A catalogue record for this book is available from the British Library.

Bibliographic information published by the Deutsche Nationalbibliothek
The Deutsche Nationalbibliothek lists this publication in the Deutsche Nationalbibliografie; detailed bibliographic data are available on the Internet at <http://dnb.d-nb.de>.

© 2017 Wiley-VCH Verlag GmbH & Co. KGaA, Boschstr. 12, 69469 Weinheim, Germany

All rights reserved (including those of translation into other languages). No part of this book may be reproduced in any form – by photoprinting, microfilm, or any other means – nor transmitted or translated into a machine language without written permission from the publishers. Registered names, trademarks, etc. used in this book, even when not specifically marked as such, are not to be considered unprotected by law.

Print ISBN: 978-3-527-34158-0
ePDF ISBN: 978-3-527-80198-5
ePub ISBN: 978-3-527-80196-1
Mobi ISBN: 978-3-527-80197-8
oBook ISBN: 978-3-527-80195-4

Cover Design Bluesea Design, McLeese Lake, Canada
Typesetting SPi Global, Chennai, India
Printing and Binding Markono Print Media Pte Ltd, Singapore
Printed on acid-free paper

This book is dedicated to my family, near and far away.

Contents

Foreword *xiii*
Preface *xv*
Acknowledgments *xvii*

1	**Introduction** *1*	
1.1	A Brief Story of Laboratory Automation *1*	
1.2	Approaches for Instrument Integration *2*	
1.2.1	The Usual Approach for Instrument Integration *2*	
1.2.2	Instrument Integration with Scripting *3*	
1.3	Scripting versus Standardization in Laboratory Automation *3*	
1.4	Topics Covered in this Book *5*	
1.5	Learning by Doing: FACACO and FAKAS *7*	
1.6	Summary *10*	
	Suggested Reading *10*	
2	**The Very Basics of AutoIt** *13*	
2.1	What Is AutoIt? *13*	
2.2	Alternatives to AutoIt *14*	
2.3	Getting AutoIt *15*	
2.4	Writing Your First Script (Mouse Click Automation) *15*	
2.5	Knowing More about SciTE *16*	
2.5.1	Writing Aids *17*	
2.5.2	The Console *18*	
2.6	AutoIt on Linux *18*	
2.7	Summary *18*	
	Suggested Reading *19*	
3	**Timed Scripts** *21*	
3.1	Controlling the Timing of Actions *21*	
3.2	Moving and Activating Windows *22*	
3.3	Sending Keyboard Inputs *23*	
3.4	"For" Loops and Variables *23*	
3.4.1	Automating FAKAS *25*	
3.4.2	First view of AutoIt v3 Windows Info (AWI) *26*	
3.4.3	AU3Recorder *28*	

3.4.4	Automating FACACO	29
3.5	Organizing Your Code: Functions and Libraries	29
3.5.1	Calling Functions from Different Files	31
3.6	Replacing Mouse Clicks with Keyboard Shortcuts	32
3.7	Summary	34

4 Interactive Scripting 35
- 4.1 Window Monitoring 35
- 4.2 Pixel Monitoring 37
- 4.3 "While … WEnd" Loops for Pixel Monitoring 39
- 4.4 Synchronizing FACACO and KAKAS Using Pixel Monitoring 40
- 4.5 Enhanced Pixel Monitoring Using PixelCheckSum 43
- 4.6 Blocking Access to Keyboard and Mouse 46
- 4.7 Summary 46

5 Scripting with Controls 49
- 5.1 Using AWI to Get Control Information 49
- 5.2 Functions That Provide Control Information 51
- 5.3 Sending Commands to Controls 52
- 5.4 Synchronizing FACACO and FAKAS Using Controls 52
- 5.5 Dealing with Errors: If … Then 55
- 5.6 Infinite Loops and Controls 57
- 5.7 Summary 59

6 E-mail and Phone Alarms 61
- 6.1 E-mail Alarms 61
- 6.1.1 Sending E-mail Using Third-Party Software 61
- 6.1.2 Sending E-mail Using SMTP 63
- 6.2 SMS and Phone Call Alarms 65
- 6.2.1 Sending SMS 65
- 6.2.2 Making Phone Calls 66
- 6.3 Summary 69

7 Using Low-Cost Equipment for Laboratory Automation 71
- 7.1 G-Code Devices 71
- 7.1.1 CNC Routers 71
- 7.1.2 G-Code for CNC 73
- 7.1.3 Synchronizing a CNC Router to a Laboratory Instrument 74
- 7.1.4 3D Printers 75
- 7.2 Robotic Arms 76
- 7.3 Do-It-Yourself Devices 77
- 7.4 Summary 77
- Suggested Reading 78

8 Arrays and Strings 79
- 8.1 Organized Data: Arrays 79
- 8.2 Raw Data: Strings 80
- 8.3 Summary 82

Contents | ix

9	**Data Processing with Spreadsheets** 83
9.1	Exporting Results to Spreadsheet Software 83
9.1.1	Selecting Spreadsheet Software 83
9.1.2	Transferring Data to Spreadsheets 84
9.1.3	Transferring Data in Real Time 87
9.2	Dealing with Saved Results (Files) 87
9.3	Processing Spreadsheet Files 91
9.4	Summary 94

10	**Working with Databases** 95
10.1	Starting SQlite in AutoIt 95
10.2	Creating SQlite Databases 96
10.3	Modifying an Existing SQlite Database 99
10.4	Databases with More Than One Table 101
10.5	Retrieving Data from Databases 102
10.6	Summary 104

11	**Simple Remote Synchronization** 107
11.1	Time Macros 107
11.2	Synchronizing FACACO and FAKAS Using Time Macros 108
11.3	Summary 109

12	**Remote Synchronization Using Remote Control Software** 111
12.1	TeamViewer 111
12.2	Synchronizing FACACO and FAKAS Using TeamViewer 112
12.3	Summary 115

13	**Text-Based Remote Synchronization** 117
13.1	Choosing Instant Messaging Software 117
13.2	Writing and Reading from Trillian Using AutoIt 119
13.3	Synchronizing FACACO and FAKAS Using Trillian 121
13.4	Summary 123

14	**Remote Synchronization Using IRC** 125
14.1	AutoIt and IRC 125
14.2	Monitoring the Connection 126
14.3	Synchronizing FACACO and FAKAS 130
14.4	Final Considerations 132
14.5	Summary 133

15	**Remote Synchronization Using Windows LAN Tools** 135
15.1	Connecting to a LAN 135
15.2	Creating a Shared Folder 137
15.3	Synchronizing FACACO and FAKAS 139
15.4	Summary 140

16 Remote Synchronization Using Third-Party LAN Software 143
16.1 Connecting to a LAN Using Bingo's Chat 143
16.2 Automated Communication Using Bingo's Chat 144
16.3 Synchronizing FACACO and FAKAS 147
16.4 Summary 148

17 Interacting with Devices via COM Ports 149
17.1 Serial Communication Protocols 149
17.2 AutoIt and COM Ports 150
17.3 Monitoring in Real Time 153
17.4 Implications for Other Devices 157
17.5 Other Technologies for Instrument Control 157
17.6 Summary 157
Suggested Reading 158

18 Introduction to Graphical User Interface (GUI) 159
18.1 Making a Very Simple GUI 159
18.2 Adding Simple Elements to a GUI 161
18.3 Setting Keyboard Shortcuts 163
18.4 Summary 165

19 Using GUI to Control Instruments 167
19.1 GUIs to Control the EHMA Valve Actuator 167
19.2 Controlling Two or More COM Ports in the Same Script 169
19.3 A GUI to Control a Digital Balance 171
19.4 Summary 174

20 Multitasking GUIs 177
20.1 The "GUIOnEventMode" Option 177
20.2 Multitasking Using GUIOnEventMode 179
20.3 Summary 182

21 Adding Graphical Elements to a GUI 183
21.1 Getting Started with GDIplus 183
21.2 Creating Animations Using GDIplus 185
21.3 Summary 189

22 Creating GUIs Using Koda 191
22.1 Getting Started with Koda 191
22.2 Creating a Script 194
22.3 Summary 196

23 Some Suggestions 197
23.1 For Manufacturers: All Instruments with a GUI 197
23.2 For Manufacturers: All GUIs with Access to Controls 197
23.3 For Manufacturers: Stop Developing Standards for Laboratory Automation 197

23.4	For Users: Hardware Trumps Software	*198*
23.5	For Users: If You Can, Choose Controls	*198*
23.6	For Users: AutoIt May Not be the Best Programming Option in Some Cases	*198*
23.7	For Users: Be Aware of Technological Advances	*199*
23.8	For Users and Manufacturers: AutoIt Scripts May Serve as Basis for New Products	*199*
	Suggested Reading	*199*

A	**Other SciTE Features**	*201*
A.1	Code Wizard	*201*
A.2	Organizing Your Scripts with Tidy	*202*
A.3	Tools that Facilitate Navigation	*203*

B	**Optical Character Recognition**	*207*
B.1	OCR in AutoIt	*207*
B.2	Copying from the Screen and Applying OCR	*209*

C	**Scripting with Nonstandard Controls (UIA)**	*211*
C.1	Downloading the UIA Software Package	*211*
C.2	Sending Instructions	*212*
C.2.1	Mouse Clicks	*213*
C.2.2	Keyboard Inputs	*216*
C.3	Getting Information about Controls	*217*
C.3.1	Getting Information from FAKAS Controls	*218*
C.3.2	Getting Information from Controls of Other Programs	*220*
C.4	Automating a LabView Program	*221*
C.5	Summary	*222*

Index *223*

Foreword

Would you like to have a lab assistant willing to do the most tedious repetitive tasks in your laboratory perfectly every time? Would you like that lab assistant to work for a few cents an hour for 24 h a day, 7 days a week, 365 days a year? Better yet, would you like the salary of that lab assistant covered under the operational overhead of your institution?

This is all more than possible with the laboratory automation wizardry explained in detail by Matheus Carvalho in this book. Practical Laboratory Automation – Made Easy with AutoIt is a tour de force in laboratory automation. It allows scientists to harness automation to radically increase productivity as is common in industry. Specifically, Carvalho guides even the most inexperienced user into a deep dive of AutoIt in the context of laboratory automation. AutoIt is a zero-cost software tool that automates mouse clicks and keyboard entries in Windows. These properties enable seamless integration into the operation of other programs – like the shockingly expensive proprietary programs that come bundled with most conventional modern scientific equipment. AutoIt source code was openly released under the GNU GPL (up to v.3.2 2006), and then went on two paths. The AutoHotkey project remains open source and AutoIt became closed source, but remains freeware with a supportive online community.

My own introduction to AutoIt came several years ago when we needed to make the same measurement in my laboratory every few minutes for hundreds of hours continuously. This was a waste of time for even the least-skilled and poorly paid student researcher. The obvious solution was to add a single loop to the equipment vendor's proprietary software running the tool to make the measurements. The vendor tried to extort thousands of dollars from us to include this new functionality. Unfortunately for the vendor, a student found AutoIt. Our problem was solved in seconds. Now a few pennies of electricity would enable the worst measurement in my laboratory to be completed without any wasted student hour. Besides teaching me the value of AutoIt, this experience set me on the path of first making our own scientific tools and eventually into the open-source scientific hardware community. This has saved my laboratory hundreds of thousands of dollars and Practical Laboratory Automation may do the same for you.

Now our initial use of AutoIt is trivial compared to the super user abilities that Carvalho demonstrates throughout this book. This is a book you should read with a computer on your lap so that you can ensure you learn the skills by trying them out as Carvalho explains new techniques. You will walk away after reading it with the technical skills to automate even the most complex scientific equipment. I should point out that although the techniques Carvalho explains in this book are complex (e.g., pixel monitoring, scripting, and remote synchronization) this book is written for the nonspecialist Windows user. It really has been "made easy." Anyone who works in a laboratory – even if he/she does not know anything about coding, robotics, embedded systems, or microcontrollers will have no problem using the techniques discussed to automate his/her own equipment. As Practical Laboratory Automation correctly points out, this will enable you freedom from lock-in with the equipment vendors using Windows as the platform of their machine, which alone should save you money and help your lab function more efficiently. It also offers the potential to "resurrect" formally quasi-functional tools collecting dust in the corner of your laboratory now (such as those where the software is no longer supported). However, even more exciting is the opportunities this book gives you to use automation to tackle new types of experiments that would simply have been too time-consuming or cost-prohibitive for all but the most wealthy laboratories.

You have made a good decision to purchase this book. It will make your laboratory more efficient and save you money. My hope is that it functions as a good "gateway" drug that it pulls you into open-source scientific hardware community. This community is building from scratch progressively more sophisticated automated research tools for a tiny fraction of the cost of proprietary equipment. Both the source code for the hardware and the software are being shared openly enabling rapid diffusion and technical evolution. As you become familiar with automation and comfortable with the ability to improve the functionality of your equipment, the technical jump to hacking open hardware becomes far less intimidating. So, fire up your computer, turn the page, and get started.

Joshua Pearce
Michigan Tech Open Sustainability Technology Lab
Open-Source Lab: How to Build Your Own Hardware and Reduce Research Costs
Houghton, MI

Preface

Some years ago, I found by chance a very cheap robotic arm (~US$ 100) for sale. This robotic arm could be controlled using a computer, similarly to autosamplers of analytical equipment. A device with that capability and costing so little was a surprise to me, because I was used to autosamplers costing as much as a brand new car. At that point, I was familiar with AutoIt, the scripting language presented in this book, and I realized that using it I could try to use the robotic arm as a substitute for the expensive autosamplers. I did it, and it worked perfectly.

The test with the robotic arm made clear for me that AutoIt could be used to integrate any type of devices in the laboratory, enabling not only money saving, but also total freedom regarding the choice of equipment. As I came to realize, my colleagues were not aware of such possibilities. Therefore, I decided to share this idea with other professionals, and published two articles in scientific journals about AutoIt used in the context of laboratory automation. However, scientific articles are short, not allowing for an in-depth coverage of AutoIt, and thus I decided that a book would be a better way to disseminate the idea among laboratory technicians and scientists.

This book covers several aspects of AutoIt in the context of laboratory automation. I expect it to empower readers in two ways:

1) By providing independence from instrument manufacturers. In the same way I could use a cheap robotic arm as an autosampler for analytical equipment, readers will become able to combine any type of equipment very easily. Doing so, it will be possible for them to choose the device that is the best or the most cost effective for a given task, regardless of what manufacturers could have originally suggested. If you are familiar with the field of laboratory automation, this may come as a surprise. Nowadays, the prevalent view is that integration of laboratory equipment is very complex and accessible only for specialized professionals. This book will demonstrate that with AutoIt such integration is very simple and accessible to virtually any laboratory technician or scientist.

2) By presenting AutoIt. AutoIt is not just another scripting language. Because it automates mouse clicks and keyboard entries, AutoIt is a very powerful tool that allows complete automation of a program, and also total integration among different programs. Being free to get and easy to learn, AutoIt is a fantastic tool that can be used not only in the context of laboratory automation, but also as a general way to easily increase productivity when using the computer.

AutoIt is a scripting language, and scripting is a type of programming. Although many readers probably have some background in programming, I wrote this book assuming that the reader has only a little or no such background. I did so to make the book accessible for all interested readers, in such a way that they do not need to first learn to program to only then learn AutoIt. AutoIt is an ideal first programming language, both for its simplicity and practical scope. In fact, it could be said that AutoIt is fun, because there is an inherent satisfaction when solving problems, especially difficult problems. AutoIt makes it easy to solve some usually perceived as difficult problems.

I hope you find this book useful, and that you can apply some of the techniques here presented in your work. AutoIt has not only made part of my work much easier, it has also enabled substantial economic savings in the operation of the laboratory. I am sure that the same can be true for you.

23 July 2016 *Matheus Carvalho*
Lismore, NSW, Australia

Acknowledgments

I am very grateful to Alessandra Carvalho, who helped me make the book more readable, and Douglas Tait and Joshua Pearce, who made useful comments on the book.

I am also grateful to Reinhold Weber, from Wiley VCH, who believed in the idea of the book and helped make it become real.

1

Introduction

1.1 A Brief Story of Laboratory Automation

Through human history, there has always been a quest for automation. For instance, the water mill, which has been used by different ancient cultures, can be considered an example of mechanical automation in the general sense of replacing human labor with a more reliable and powerful alternative. By replacing human labor with a water mill, our ancestors achieved the following aims of automation: reduced production costs, increased production efficiency, and improved safety in the production process. With the industrial revolution, when powerful engines became available, automation in this general sense maximized production efficiency as never before in human history. Further development of electronics and computing brought automation to its present and familiar stage, in which finely controlled motors execute precise tasks emulating (and surpassing) human efforts, movement sensors open doors for people to pass through, and computer programs save works without users needing to remember doing that.

Scientific laboratories have adopted automation as its technology developed. The first documented solutions in laboratory automation were devised by scientists to improve their own work. Devices such as automated filters and siphons have been built by ingenious means since the late nineteenth century. With the advent of electronics, a wide range of new devices, such as conductivity meters, gas analyzers, pH meters, and automated titrators, became available. Soon after the Second World War, automation tended to become predominantly provided by specialized companies making the devices, due to the increased complexity in manufacturing. The exponential progress in computing then enabled the opening of the first fully automated laboratory by Dr Masahide Sasaki at the Kochi Medical School in Japan in the early 1980s. In the following decade, similar laboratories were opened in Japan, the United States, and Europe. The approach followed in such fully automated laboratories started becoming known as *total laboratory automation* (TLA).

TLA was and still is very expensive. Because of that, only a few laboratories, normally those involved in fields that can produce high financial returns, such as drug discovery, can afford to implement it. In order to make TLA more accessible for medium- and small-scale laboratories, the concept of modular automation was introduced in the late 1990s. Through this concept, smaller laboratories could purchase one or a few automated instruments and progressively upgrade them when money became available. A very recent development in modular automation that has the potential to reduce even further costs in laboratory automation is the adoption of open-source hardware, which has its blueprints freely available. Enabled by open-source microcontrollers, and open-source building devices such as three-dimensional (3D) printers, this technology enables users to build their own devices at a surprisingly low cost.

Modular automation, including open-source hardware, however, does not work in all situations because of the difficulty in integrating instruments built by different manufacturers. The negative effect of this lack of integration cannot be overemphasized. It has been recognized as a problem for more than 25 years, and attempts have been made to resolve it by means of standardization of the communication between instruments. Unfortunately, such standardization has never become widespread, and to date, it has remained difficult to integrate instruments built by different manufacturers. This book aims to present a way to overcome this limitation, and thus enable laboratories to implement automation at a much lower cost than by traditional means. The approach presented in this book is very simple and accessible for most professionals in laboratories even if they do not have a background in electronics or computing. It is also a very powerful approach, which overcomes virtually any lack of compatibility between instruments.

1.2 Approaches for Instrument Integration

As explained above, a fundamental aspect of laboratory automation is instrument integration, that is, the ability to make instruments work together. The following are the two ways to enable instrument integration.

1.2.1 The Usual Approach for Instrument Integration

Communication between laboratory instruments is usually implemented by a computer controlling one instrument, which in turn controls others (Figure 1.1). There is data interchange between the computer and the first instrument, and between instrument 1 and the other instruments. However, there is no direct communication between the computer and the instruments being controlled by instrument 1. In other words, only instrument 1 can be directly controlled by the computer and, consequently, by the user. An advantage of this approach is simplicity: the user only needs to operate a single program that controls the whole set of instruments.

However, this approach also considerably limits options for users. For example, let us assume that instrument 2 breaks down and needs to be replaced. The user then finds an alternative to instrument 2, which performs better and costs less than the usual instrument supplied by the manufacturer of instrument 1. Ideally, the user should be able to connect this alternative instrument to instrument 1 and continue the work. In practice, however, this is not possible in most cases, because instrument 1 was built to communicate exclusively with instrument 2 and vice versa. Therefore, the user has no choice but to buy a second instrument 2.

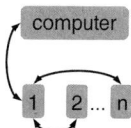

Figure 1.1 Common approach to make instruments work together in a laboratory. Arrows show the paths for data exchange.

Figure 1.2 Instruments working together in a framework enabled by scripting. As in Figure 1.1, arrows indicate exchange of data.

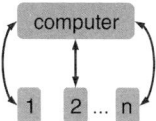

1.2.2 Instrument Integration with Scripting

The limitation of the traditional approach of instrument integration (Figure 1.1) can be eliminated by scripting. As explained above, with scripting, the user coordinates the programs that control different instruments. This way, if the different instruments set up to work together have each a software interface, they can be integrated using scripting (Figure 1.2).

In Figure 1.2, there is no direct communication between instruments; instead, the computer communicates with all instruments. It is important to note that this is conceptually different from the scenario in Figure 1.1, in which the computer controlled only instrument 1, and instrument 1 controlled the others.

In the example outlined earlier, the user needed to replace instrument 2. If the user finds a replacement for instrument 2, which has a program controlling it, he/she can readily make it work together with instrument 1 using the arrangement in Figure 1.2. To do so, the user does not need any knowledge of electronics or even advanced computing; scripting is all that is necessary. Thus, in our hypothetical history, the user could both save money and obtain a better instrument.

The example is hypothetical, but resembles the routine activities of a laboratory technician. It is common that when replacing a broken instrument the normal alternative is either too expensive or takes a long time to become available. In such cases, the damaged instruments could be replaced by cheaper and ready-to-use alternatives, by means of scripting. Two examples can be found among the reading suggestions at the end of this chapter. In one, the autosampler of a machine was coupled to a water analyzer, because the analyzer did not have an autosampler, and was not even designed to work with one. In the other case, a low-cost robotic arm (less than US$ 500) was used as autosampler for an automated titrator, rather than using the autosampler originally designed for that instrument, which would cost more than US$ 50 000.

Another aspect that gives an advantage to scripting is that it is not always possible to use one instrument to control several others (Figure 1.1). In most cases, instruments are designed to communicate with only another one. With scripting, there is no limit on the number of instruments that can be synchronized.

1.3 Scripting versus Standardization in Laboratory Automation

Scripting is not the first proposed solution for the problem of lack of compatibility between instruments. However, as will be explained below, it is the only one with real odds of solving this problem.

The usually proposed solution for the compatibility problem in laboratory automation is the adoption of standards. Such standards would mean that all instruments would communicate using the same protocols. This would make the integration of instruments very simple, and the problem of replacing instrument 2 (Figure 1.1) illustrated above would be very easily solved.

Although such a solution looks desirable, it is very difficult to be put in practice. A recent effort in this direction that stands out by its large magnitude is SiLA (Standards in Laboratory Automation). According to their website, http://www.sila-standard.org, SiLA is a consortium of several system manufacturers, software suppliers, system integrators, and pharmaceutical corporations, among others. There are several working groups composed of high-skilled experts dealing with device control and interfaces, command dictionary specification, data standard, process management system, and so on. If, on the one hand, it is impressive to see such a combined work with the aim of improving laboratory automation, then, on the other hand, the necessity of such endeavor warns that the problem at hand is complex.

In addition to the complexity of the problem itself, there is resistance of manufacturers in adopting a potential standard that may or may not become widespread. As the proposers of SiLA themselves admit (see the relevant publication in the reference list at the end of this chapter), newcomers need to spend time, resources, and money to implement SiLA. In addition, SiLA will be successful only if all players in the industry adopt it. It is a very ambitious goal, which is still far from being reached.

Now let us compare this situation with scripting. The first aspect is that the technology for enabling scripting is ready; there is no need for further development (although of course this development continues in the form of bug fixing and improvements in the scripting language). In other words, integration of laboratory instruments using scripting can be implemented now, while uniform standards are not yet available for most laboratory equipment.

The second aspect is that it is much easier for manufacturers to adapt to scripting than to adopt standards. In fact, manufacturers do not need to do anything, as scripting was designed to deal with software "as is." As will be explained in the last chapter of this book, however, there are measures that manufacturers can adopt if they want their software to be "script-friendly." These are minor modifications that are much easier to adopt than standards.

The third aspect is that standardization necessarily creates restrictions to further development. If a new and more efficient way of implementing a software solution is devised, but it is outside the scope of the standard, either the solution must not be implemented or the standard needs to be modified. With scripting, there is no risk of such problem unless a very radical change in the way computers work comes to place, like, for example, the abolition of mouse and keyboard use (which is extremely unlikely).

The fourth aspect is how users need to adapt to changes brought by scripting or standards. In order for them to work, standards need a universal interface, called *integrator*, that controls all programs. Scripting also demands that users learn how to control several different programs using a common framework. This book provides information on this aspect, which, as will be seen, is quite accessible. An important point is that the programs used for scripting are free of cost. By contrast, integrators are not always free.

The fifth aspect is that adoption of standards indicates that in many cases perfectly working devices would become obsolete for just not following the standard. This could be translated into huge costs of modifying existing equipment, or obtaining new ones. Nothing of this is necessary if scripting is used, since it demands no modification in the equipment. Another consequence of this aspect is that old equipment, many times

Table 1.1 Summary of differences between scripting and standards for laboratory automation.

	Scripting	Standards
Technological availability	Ready	In development
Backward compatibility	Yes	No
Universal acceptance	Not necessary	Necessary
Difficulty for manufacturers	None	High
Easily adaptable for new technologies	Yes	No
Difficulty for users	Little	Little to high
Cost for manufacturers	Zero or small	High
Cost for users	Zero or small	High in most cases

compatible with only older computers, can be "resurrected" and made to work with newer counterparts (up to a point; Windows versions before XP are not fully supported, see more details in Chapter 2).

The sixth aspect is that by enforcing standards, laboratories miss the opportunity of using the myriad of new devices and technologies that appear at a frantic pass every day. These devices do not necessarily follow the standards that are proposed to be used by laboratory instruments. Therefore, any creative use of traditional instruments and revolutionary new devices becomes hampered with the adoption of standards. The same is not true with scripting: any new device can be used in conjunction with traditional laboratory instruments by means of this technology.

In summary, scripting is by definition a better solution than the adoption of standards for laboratory automation. It is easier, cheaper, and can be implemented now. Standards cost money, time, and are not ready yet (and may never be). Table 1.1 summarizes the differences between the two approaches.

1.4 Topics Covered in this Book

Before knowing what this book is about, it may be important to clarify what it is not about, because this book differs in many aspects from previously published books on laboratory automation.

The previously published books on laboratory automation often covered electronics, and even some aspects of mechanics. They also often presented deep discussions about communication protocols, or some advanced concepts in programming. In some sense, these books demanded (or aimed to provide) an encyclopedic knowledge to users who are already necessarily specialized in another profession. This book is different from these previous ones in that it does not cover any of those diverse subjects, except superficially for some of them. This is because of the novel approach presented here: instead of building instruments or creating software interfaces for them, using scripting we can simply coordinate the software interfaces, which are already easily available for existing instruments and combine them. Therefore, this book is mainly focused on how to use scripting to coordinate software interfaces, which is much more accessible than the subject of previously published books for the ordinary laboratory technician or scientist without a deep background in computer science or robotics, and without time to devote to learning a complex new subject.

It is important that the book be read in sequence, especially for those without knowledge on programming. It is suggested for those who do not have a background in programming to read this book from Chapters 1 to 5, at least, and test all the scripts presented therein. The next chapters can be understood much more easily, provided the initial ones are studied first. For readers with experience in programming, it can still be useful to have a look into the initial chapters to know how AutoIt works. It will be seen that the codes used throughout this book are often very simple, and the readers will probably have no problems understanding them. A reference for the language is not presented in this book, rather only the aspects of the language that are useful in specific contexts are presented. Therefore, chapters about loops or conditionals will not be found; however, these subjects are presented as parts of chapters dealing with their uses, as they become necessary. A final comment for those with a background in programming: throughout the book, the "best practices" for writing scripts in AutoIt are deliberately ignored, but are available at http://https://www.autoitscript.com/wiki/Best_coding_practices. It has been done to simplify the learning for people without a background in programming, as these best practices may be confusing for uninitiated readers.

Chapter 2 introduces AutoIt. As it is explained there in detail, AutoIt stands out as a very powerful yet simple scripting language that is perfect to be used for laboratory automation.

Chapters 3–5 are arguably the most important in this book. They describe how AutoIt can be used to synchronize instruments built by different manufacturers and controlled by a single computer. All examples in these chapters are for the teaching software written for this book.

Chapter 6 shows how AutoIt can be used to implement alarms to deal with problems during automated measurements. In this case, not only the teaching programs are used, but also third-party software is introduced, demonstrating that AutoIt is versatile and can combine communication software very easily with laboratory instruments.

Chapter 7 shows how AutoIt can be used to integrate low-cost automation devices, such as CNC routers, 3D printers, and robotic arms, with laboratory equipment. Again, third-party software is also used in the examples.

Chapter 8 was written for readers without any background in programming, as it covers arrays and strings. However, readers with some programming background can also find this chapter beneficial by reading it to check the details of the language on these topics.

Chapters 9 and 10 show how AutoIt can automate the export of data generated by instrument software, and how the processing of such data can be made easier and faster.

Chapters 11–16 present several techniques that enable the synchronized control of instruments by more than one computer. In most cases, this demands the use of computer networks, including the Internet. Most are very simple, but Chapter 14 demands deeper study.

Chapters 17–22 cover how AutoIt can be used as a programming language and not a scripting language. In other words, AutoIt is used to directly control an instrument, by means of building user interfaces and using communication protocols. These chapters are more advanced than the rest of the book, but still accessible for readers with a background in programming or those who follow the book chapters in order from the beginning.

Chapter 23, the last one, presents some suggestions and guidelines to users and manufacturers of laboratory instruments so that both sides can take advantage of scripting. It is solely a personal view, but hopefully it will help to convince instrument manufacturers to implement a few small changes that can make laboratory automation even more accessible for laboratory users, and orient users to choose the best options when implementing laboratory automation.

After the main text, there are three appendices in this book. The first one shows some useful features of the standard script editor for AutoIt, which are not essential but can help the user in the day-to-day writing of scripts. The second appendix presents optical character recognition (OCR), a technology that is still imperfect, but could be very useful for scripting in some situations. The third one also presents a developing technology, which is the use of Windows User Interface Automation in AutoIt. This technology in particular can be very useful for automating some popular programs used in the laboratory, which are those developed using LabView.

1.5 Learning by Doing: FACACO and FAKAS

An aspect of this book is that it is entirely based on examples, because the author believes that the best way to learn to script is by practicing. Scripting is so simple that in most occasions readers will be able to simply copy the examples, and, with minimal modifications, have a script working for their needs. Still, in order that the examples are fully understandable, it is necessary that readers gain access to the same software than the author. In order to ensure that, two very simple programs that share similarities with software normally used for controlling laboratory equipment were written: FACACO and FAKAS. FACACO stands for FAke Carbon Analyzer Controller, and FAKAS FAKe AutoSampler.

The first step in the learning process is to become familiar with the two programs. They will be described in detail, but it could be seen that a fundamental requisite when scripting is to be very familiar with the programs being automated.

First, download them from http://www.wiley-vch.de/publish/en/books/ISBN978-3-527-34158-0/. Warning: The code for these two programs was written using AutoIt, and some antivirus software treated them as contaminated with a virus. FACACO and FAKAS are safe to use. The programs do not need to be installed. Double click each of the icons to see the interfaces.

FACACO simulates the controller of a carbon analyzer for water samples. Without going deep into this type of analysis, it is useful to understand the fundamentals of it in order that FACACO becomes more familiar. There are two different types of dissolved carbon in water: dissolved inorganic carbon (DIC) and dissolved organic carbon (DOC). FACACO simulates an instrument that pumps a fixed amount of a water sample to a reaction chamber where DIC and DOC are extracted. DIC is extracted by acidification, while DOC is extracted by oxidation. Both extractions consist in converting DIC or DOC to CO_2. The extracted CO_2 is then measured using an appropriate sensor. A relevant detail is that DIC must be extracted beforehand so that DOC can be measured in isolation.

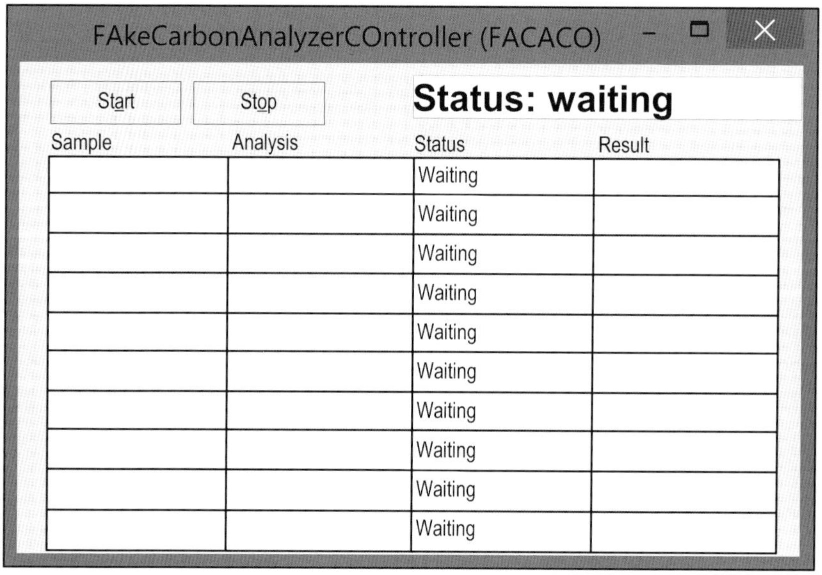

Figure 1.3 FACACO interface.

FACACO interface contains two buttons: Start and Stop (Figure 1.3). There is a spreadsheet-like table below it, with four columns: Sample, Analysis, Status, and Result. Finally, there is the phrase "Status: waiting" in large font inside a white background.

FACACO can be explored by clicking on the buttons and typing into the cells. For example, when you press the Start button, you will see a pop-up message asking if you really want to start the measurements, and you can say yes or no. If your "Sample" cells are all empty, nothing will happen, regardless if you press yes or no. Then, let us fill the "Sample" columns with some names, like "sample1" or "sample2,". Now try the Start button once more, and respond yes to the pop-up box. You will see another pop-up box complaining that the type of analysis must be DIC or DOC. Then, on the "Analysis" column, fill the analysis type (DIC or DOC) for each sample, like, for example, in Figure 1.4.

Now, press Start again and say yes to the pup-op window. Another pop-up window will appear, this time with a warning that the sample must be connected to the pump (the program assumes that there is no autosampler and that the user will replace the vials connected to the pump through the sampling line). After that, the sample will be measured, and you will see the contents of the "Status" column change. For DIC analysis, the status changes from Waiting to Sampling to Measuring DIC to Calculating DIC, and to Done. Similar changes appear in the large Status field, including changes in the color of its background. When done, the result of the analysis appears in the Result column. For DOC, the sequence of status is Waiting, Sampling, Removing DIC, Adding oxidant, Measuring DOC, Calculating DOC, and Done. The removing DIC step is to ensure that the sample will be acidified before measuring DOC, as explained before for carbon analyses in water. The pop-up window appears for the next sample, and for all subsequent samples until the end of the list.

You can press the Stop button anytime during the measurement. Doing so, the status of the sample currently being measured goes to "Interrupted" and the background

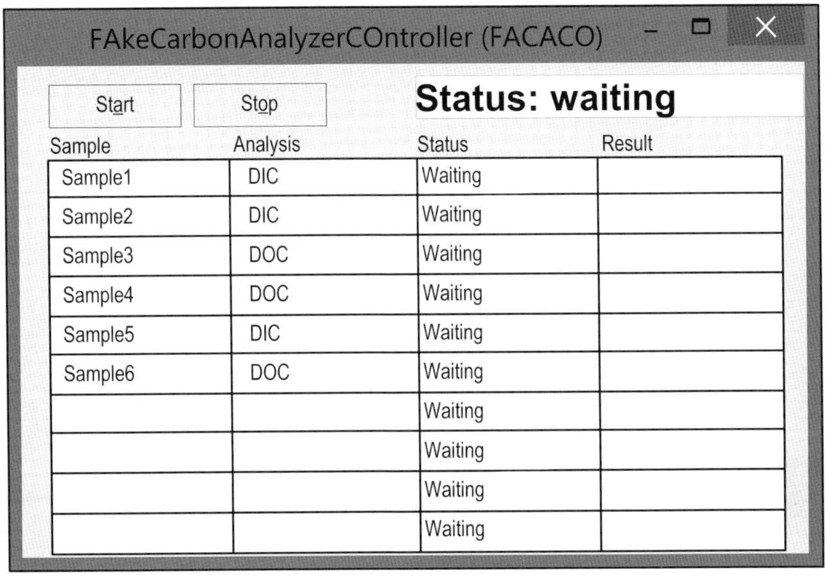

Figure 1.4 FACACO interface with samples and analyses loaded.

behind the large Status label becomes red. If you press start again, you should be able to continue with the remaining samples. If you want to start over, the easiest way is closing the program and opening it again.

FAKAS (Figure 1.5) simulates the controller of an autosampler for water samples. This imaginary autosampler can handle 10 samples, and be imagined as a carousel, where these samples stand on, and a needle connected to a tube. This tube is connected to the pump of the hypothetical carbon analyzer controlled by FACACO. Therefore, in order to sample water from a given sample, we need to tell FAKAS its position (between 1 and 10), tell it to move to the sample by pressing the button "Go to sample," and then, when there, tell it to push the needle down ("Push needle down" button) so that it reaches the

Figure 1.5 FAKAS interface.

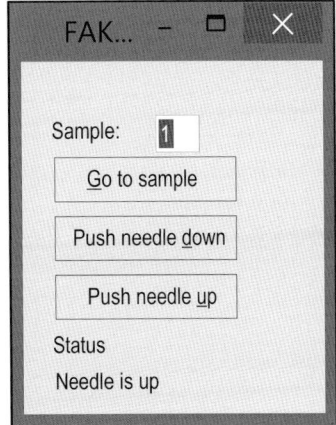

water in the sample vial. When the sampling is done, we can tell the autosampler to pull the needle back up ("Pull needle up" button) so that it can move to other samples. The status of the autosampler is given at the bottom of the FAKAS window. You can test the buttons on FAKAS interface and see the error messages that come if you try to move the needle when it is down, or if you input a number smaller than 1 or larger than 10, or even letters in the "Sample" field.

When writing scripts, it is fundamental to be familiar with the interfaces of the programs that control the instruments. Many times, this familiarity comes with the intention of performing the scripting. This will become clear in the following chapters.

1.6 Summary

- First efforts in laboratory automation were carried out by technicians and scientists themselves.
- Specialized companies started providing automated equipment in large scale after the Second World War.
- TLA was first implemented in Japan by Dr Masahide Sasaki in the early 1980s.
- TLA never became widespread due to high costs, which were partly driven by lack of compatibility between instruments from different manufacturers.
- Scripting is the easiest and cheapest way to enable compatibility between instruments.
- AutoIt is the scripting language presented in this book.
- Two programs (FACACO and FAKAS) are used as examples for the scripts in this book.

Suggested Reading

History of laboratory automation:

Boyd, J. (2002) Robotic laboratory automation. *Science*, **295**, 517–518.

Feiglin, M. and Sterling, J.D. (2008) Laboratory automation. *Nat. Rev. Drug Discovery*, 7, 625.

Felder, R.A. (1998) Modular workcells: modern methods for laboratory automation. *Clin. Chim. Acta*, **278**, 257–267.

Felder, R.A. (2006) Masahide Sasaki, MD, PhD (August 27, 1933 – September 23, 2005). *Clin. Chem.*, **52**, 791–792.

King, R.D., Rowland, J., Oliver, S.G., Young, M., Aubrey, W., Byrne, E., Liakata, M., Markham, M., Pir, P., Soldatova, L.N., Sparkes, A., Whelan, K.E., and Clare, A. (2009) The automation of science. *Science*, **324**, 85–89.

Olsen, K. (2012) The first 110 years of laboratory automation: technologies, applications, and the creative scientist. *J. Lab. Autom.*, **17**, 469–480.

Sasaki, M., Kageoka, T., Ogura, K., Kataoka, H., Ueta, T., and Sugihara, S. (1998) Total laboratory automation in Japan past, present and the future. *Clin. Chim. Acta*, **278**, 217–227.

Standards in laboratory automation:

Bär, H., Hochstrasser, R., and Papenfuß, B. (2012) SiLA: basic standards for rapid integration in laboratory automation. *J. Lab. Autom.*, **17**, 86–95.

Delaney, N.F., Rojas-Echenique, J.I., and Marx, C.J. (2012) Clarity: an open-source manager for laboratory automation. *J. Lab. Autom.*, **18**, 171–177.

Hawker, C.D. and Schlank, M.R. (2000) Development of standards for laboratory automation. *Clin. Chem.*, **46**, 746–750.

Fast technological development:

Butler, D. (2016) Tomorrow's world. *Nature*, **530**, 399–401.

Open-source hardware for laboratories:

Pearce, J.M. (2012) Building research equipment with free, open-source hardware. *Science*, **337**, 1303–1304.

Pearce, J.M. (2014a) Cut costs with open-source hardware. *Nature*, **505**, 618.

Pearce, J.M. (2014b) *Open-Source Lab: How to Build Your Own Hardware and Reduce Research Costs*, Elsevier.

Scripting in laboratory Automation

Carvalho, M.C. (2013) Integration of analytical instruments with computer scripting. *J. Lab. Autom.*, **18**, 328–333.

Carvalho, M.C. and Eyre, B.D. (2013) A low cost, easy to build, portable, and universal autosampler for liquids. *Methods Oceanogr*, **8**, 23–32.

2

The Very Basics of AutoIt

2.1 What Is AutoIt?

AutoIt is a scripting language for the Windows operating system based on the seminal BASIC programming language. AutoIt was introduced by Jonathan Bennett and collaborators in 1999. Since then, different versions of it have been released. In this chapter, we deal with AutoIt v3, which was released in 2004, and keeps being updated (the most recent update, at the time of writing this book, v3.3.14.1, was released on July 28, 2015).

Scripting is a type of programming. If you never wrote a computer program, you might not know exactly what it means. There are some technicalities, but, to put it simply, a computer program is a set of instructions that are passed to the computer and are characterized by being able to change according to the input that they receive, and also being repeated as many times as desired. Programs are written using programming languages, which can be classified as low level and high level. Low-level languages are the most difficult to use for most people: instructions in these languages deal with very technical aspects of the computer, such as memory addresses. Because of that, many instructions are necessary to be written to perform a single action. High-level languages, by contrast, summarize the actions in a single instruction that resembles "human" language much better than low-level languages. Scripting languages are high-level languages that are fundamentally based on coordinating (or, in the jargon, "gluing") actions performed by different software. The result is that it is much faster and easier to write a script (a program written using a scripting language) than a program using a low-level language. Therefore, scripting is often preferred to low-level languages for writing simple applications. For a more detailed explanation about scripting, the reader is referred to the references listed at the end of this chapter.

AutoIt can be classified as an "extreme" scripting language. AutoIt stretches the concept of gluing to its maximum. Therefore, as explained above, a single AutoIt instruction usually replaces many combined instructions written previously in other languages, each of them already being hundreds of instructions in low-level languages. This indicates that most instructions in AutoIt are very simple for the programmer, as they represent actions that are similar to human actions. An important implication of its extreme gluing approach is that AutoIt can use as instructions the solutions that have been found by other people for many problems. In other words, AutoIt can glue different software that was originally designed to do each one a different task, which may not be a trivial task, and combine them to do a third task that could be difficult to perform differently.

Practical Laboratory Automation: Made Easy with AutoIt, First Edition. Matheus C. Carvalho.
© 2017 Wiley-VCH Verlag GmbH & Co. KGaA. Published 2017 by Wiley-VCH Verlag GmbH & Co. KGaA.

Therefore, each new brilliant piece of software that is released for Windows can be easily integrated with others by means of AutoIt.

It is important to emphasize that AutoIt is not simply a scheduler. AutoIt is a full-featured structured programming language, which uses the same logic that makes programming languages like C#, Java, and Visual Basic very powerful tools. If you are not familiar with programming, it may suffice to know that structured programming enables from the simplest to the most complex set of instructions (algorithms) imaginable. This is very important, because it means that virtually any problem regarding the control of software interfaces can be effectively tackled by AutoIt. If you are one of those that may have tried to learn programming and found it too difficult, rest assured that for purposes of laboratory automation only the simplest scripts suffice for the vast majority of cases. Nevertheless, AutoIt also allows advanced programmers to develop complex applications if necessary.

2.2 Alternatives to AutoIt

You may be thinking that AutoIt is not unique in the respect of mouse click and keyboard automation, and you are right. Some scripting languages, such as VBA (Visual Basic for Applications) for Microsoft Word and Excel, for example, allow the user to record mouse clicks and keyboard inputs thus enabling the automation of repetitive tasks. However, for purposes of laboratory automation, these languages are limited, because they can only be used as a part of the main application being automated. Therefore, for example, VBA for Excel was designed to automate tasks in Excel, and it is impossible, or at least quite complicated, to integrate the actions of Excel to those of other software. AutoIt, in comparison, works with any program written for Windows.

A different type of scripting is batch programming, in which several instructions can be organized in a file, which can be run as a program so that the instructions are followed in the desired order. The old operating system from Microsoft, the Disk Operating System (DOS), allowed the use of batch files. More recently, Microsoft has released Powershell, which enables powerful capabilities in terms of scripting. However, it is better suited to other applications more related to system administration than proper automation of common tasks usually performed by users. Therefore, AutoIt is a better option to automate tasks in a laboratory context.

Finally, the real AutoIt rival: AutoHotKey (AHK, http://ahkscript.org). Similarly to AutoIt, AHK can automate mouse clicks and keyboard inputs. In fact, AHK does almost everything that AutoIt does. This is not surprising, since AHK originated from an older version of AutoIt (v2). Some people find it easier to use AHK than AutoIt. This is a matter of personal opinion; however, AutoIt has a syntax based on BASIC, which the author finds very easy to understand. AHK has a different syntax, which, from the author's viewpoint, looks less polished than that of AutoIt. Regarding complexity, as will be seen, AutoIt is used in a very simple way in this book, which should be accessible to people without a background in programming. Those with a background in programming may even find some sections trivial.

In addition to AHK, there are other programs and scripting languages that work in potentially similar ways to AutoIt. For example, Sikuli (http://www.sikuli.org) is an interesting language that uses screenshots as part of the coding. Other options can be searched on the Internet, but it is important to note that not all of them are free, as AutoIt.

It is important to mention that AutoIt can also be used for purposes similar to those of more traditional programming languages, that is, to write programs that do not have the main purpose of simply organizing the actions performed by other programs. In fact, AutoIt has an extensive list of features, and this book will explore only a few of them that are potentially more accessible and useful in the context of laboratory automation. As you progress through this book, however, you will learn several techniques that allow the development of conventional programs, as communication using COM ports and the design of graphical user interfaces.

2.3 Getting AutoIt

AutoIt runs on Windows XP, Vista, 7, 8, or 10 (at the time of writing this book). AutoIt can also run on some previous versions of Windows (95, NT4.0, and 2000), but it may not be fully functional. Therefore, you need a computer equipped with one of those versions of the Windows operating system, and be familiar with its basic operation to properly use AutoIt.

AutoIt, is freely available for download at http://www.autoitscript.com/site/autoit/downloads/. When installing the program, there will be a prompt asking whether you want the script to run or to be edited when its icon is double-clicked. The author personally prefers that the script be edited, and then run it from the editor afterward. If you choose the opposite, you can always change this later.

It is recommended to download the script editor (SciTE), also at http://www.autoitscript.com/site/autoit/downloads/. Using SciTE to write AutoIt scripts is not necessary (Notepad could be used), but as you will see SciTe is a very-well-crafted editor that helps considerably the process of writing scripts.

2.4 Writing Your First Script (Mouse Click Automation)

Probably nothing is more impressive when starting working with AutoIt than mouse automation. Therefore, let us start with it, as it is a very useful feature as well. Open SciTE and type the following line:

```
MouseClick("left",500,500) ;mouse click at position 500,500
```

Code sample 2.1 First script using mouse clicks.

Save your script and press the key F5 (or go to the "Tools" menu and select "Go"). You should see the mouse pointer moving. Congratulations, this was your first script! You probably could not see, but the mouse also clicked at the position that it stopped. Everything written after ";" are comments, and not interpreted as code by AutoIt. You can write anything as a comment, as long as you do it after a ";". It is a good idea to include many comments in your code so that you remember what you did when you need. However, comments will not be included in the codes presented here, as they have been already explained in the text, and thus comments would only take space in the examples given. A final note about comments: they can also extend to several lines, instead of only as a part of a single line. See an example:

```
#cs
The instruction below makes the mouse
to click at the position 500,500 on the desktop
#ce
MouseClick("left",500,500)
```

Code sample 2.2 Script showing comments over several lines.

The delimiters for comments spanning several lines are "#cs" to start, and "#ce" to end.
Let us remove the comments and add another line, just below the one that you wrote, so that the script becomes like this:

```
MouseClick("left",500,500)
MouseClick("left",400,10)
```

Code sample 2.3 Script setting two mouse clicks at two different positions.

Press F5 again and see what happens. You should have seen the mouse moving twice this time, the second time nearer the upper left corner of the screen. In this script, we sent an instruction twice to the computer, the "MouseClick" instruction. Inside the brackets are the particulars of the instruction: the clicks were done with the left button of the mouse, the first at the position $x = 500$ $y = 500$, and the second at $x = 400$ $y = 10$. It is important to note that the word "left" needed to be typed between " "; this is a requirement when you pass words as arguments in AutoIt. Numbers, however, did not need " ". You note, based on the behavior of the MouseClick instruction, that the coordinate system adopts $x = 0$ and $y = 0$ as the top left corner, and that x is the horizontal axis, and y the vertical one. A different coordinate system can be selected based on the corner of a window, for example, but there is not much use for it on most occasions.

AutoIt has many instructions like MouseClick. The proper name for them is "functions." Functions are blocks of instructions that perform a well-defined task. After reading Chapter 3, you can write your own functions. Functions often receive arguments, as we saw for MouseClick. It is fundamental to pass the correct arguments so that functions perform their tasks correctly.

2.5 Knowing More about SciTE

SciTE is a very-well-crafted editor for AutoIt full of useful features. Two types of features that are very helpful have been listed here, but many of others have been left to show when they become useful in the context of the work being done in the following chapters. In addition to them, other SciTE utilities are listed in Appendix A.

2.5.1 Writing Aids

When writing scripts in the previous examples, you may have noted that as you typed "MouseClick" SciTE a small menu appeared with a list of possible functions that you could wish to write (Figure 2.1). This is one of the SciTe features, which is very useful because it helps to avoid misspelling.

You probably also noted that, when you typed "MouseClick," instructions for passing the arguments to the function appeared under the writing line (Figure 2.2), as well as a one-line summary of the function capability. Thus, it is not necessary to remember all arguments for all functions, nor what the functions do. This is definitely a great aid when writing AutoIt scripts using SciTE.

Let us have a deeper look at MouseClick, based on the description in Figure 2.2. The first argument is "button," which we learned we should replace with "left" if we want a click from the left mouse button. Then, comes a "[". This means that the argument is optional. Therefore, you can make MouseClick work even without giving it coordinates to click; it will simply click where the mouse pointer is sitting at the moment (try it!). You also do not need to specify that you want one click, this is the default (indicated by clicks = 1 in Figure 2.2). However, if you want to have more than one click, you should write the number. Finally, you can even control the click speed, which here is indicated by 10, the standard value, but that could be any number between 1 (slowest) and 100 (fastest). Throughout this book, many of these optional parameters will be ignored, as they are often unnecessary.

Once you have a file that is saved, for example, when you tested Code sample 2.1, you can have even further help with your functions. Just type the full name of the function (you can, of course, use the help of the suggestions given as in Figure 2.1), and then press F1. The help file page for the function appears with a full description of the features of the function. The help file is very well structured and is "the" reference for all functions in AutoIt. However, learning AutoIt from the help file is not easy, even with the examples provided there, unless you are an experienced Windows programmer.

In addition to helping with the function to be used, SciTE is full of features to help write scripts for AutoIt. A nice one is that it uses different colors for different elements in the text. Therefore, after some practice, you can know if something is wrong by just looking at it.

Figure 2.1 SciTe autocompleting the typing of a function.

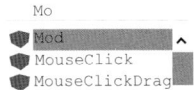

```
MouseClick(
MouseClick("button"[,x,y[, clicks = 1 [, speed = 10]] )
   Perform a mouse click operation
```

Figure 2.2 SciTe suggesting arguments to a function.

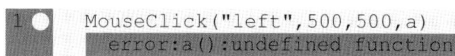

Figure 2.3 SciTE indicating that there is an error. A yellow circle appears at the left of the line with the wrong input, a red line is inserted below the line with the wrong input, and some extra comments are shown in the console (not all shown in the figure).

2.5.2 The Console

SciTE also opens a large "status bar" at the bottom of the window when the script is being run. This "status bar" is the console. There you find, even for the simplest script, a wealth of technical information, including the location of the script file in the disk and the duration of the script run. Among these, an important one shows that "AU3Check" did a test of the script. This means that the script is only run if it is syntactically correct. This is a very useful feature of SciTE, and if there is a syntactic error in the script, an error message pops up asking if you want to continue running it or not. You should say "No" and check the code; the problematic part (or parts) will be highlighted with the probable reason for the error (Figure 2.3). If you just want to know if your script is correct, but do not want to really make it work, instead of pressing F5 you can press Ctrl + F5, and the error messages will be shown as well, allowing you to correct the mistakes.

The console also teaches that the script can be stopped at any time by pressing Ctrl + Break, or restarted by pressing Ctrl + Alt + Break. If your keyboard does not have the Break key, you can choose Tools menu from SciTE and select "Stop Executing." This feature is more useful for scripts that take a long time to finish, or that never finish (as in infinite loops, which will be introduced later).

2.6 AutoIt on Linux

It is possible to use AutoIt on Linux by means of Wine, which is the software that emulates Windows on Linux. The main advantage of Linux is that, contrary to Windows, it is free. However, AutoIt does not work 100% the same on Linux as on Windows. Furthermore, only Windows applications running on Wine can be automated using AutoIt; those for Linux will be out of reach. Therefore, unless you really cannot afford to buy the Windows operating system, or if it is for some reason forbidden in your workplace (maybe a radical open-source environment), I do not suggest using AutoIt on Linux.

2.7 Summary

- AutoIt is a scripting language for Windows based on BASIC.
- A main feature of scripting is "gluing," which means combining the actions of different programs.
- AutoIt enables gluing by means of automation of mouse clicks and keyboard entries, clearly a very accessible approach for most computer users.
- Currently, the only real alternative to AutoIt in its niche is AHK.
- AutoIt and its code editor, SciTE, can be downloaded for free at http://www.autoitscript.com/site/autoit/downloads/.

- MouseClick is a function that allows you to control mouse clicks.
- SciTE has several useful features, such as autocomplete, explanations about functions as they are typed, and a console at the bottom of the screen, which shows information about the code.

Suggested Reading

Barron, D.W. (1999) *The World of Scripting Languages*, John Wiley & Sons, Inc.

Ousterhout, J.K. (1998) Scripting: higher level programming for the 21st century. *IEEE Comput.*, **31**, 23–30.

3

Timed Scripts

In Chapter 2, you were introduced to the absolute minimum to start using AutoIt. In this chapter, we will progress through important features of AutoIt and demonstrate that with only a little more knowledge we can write useful scripts that enable one type of laboratory automation: the synchronization of instruments from different manufacturers. As explained in Chapter 1, this is not trivial in the common laboratory. You will see how easily this can be accomplished with AutoIt.

In this chapter, the simplest type of scripting for synchronizing laboratory instruments will be presented: scripts based on timing. With this type of scripts, actions are intercalated with carefully determined timing to enable two or more instruments to be synchronized correctly.

This chapter can be considered the most important throughout this book. In fact, by studying only this chapter, you should already become able to use AutoIt in a productive way for laboratory automation. In addition, several subsequent chapters will be based on it. Therefore, take your time to learn well its contents. And do not worry, it is quite simple.

3.1 Controlling the Timing of Actions

Check Code sample 2.2. There, we had two lines, each for a mouse click. Let us increase our degree of control by determining not only where the mouse pointer clicks, but also when. To do so, we use the "Sleep" function, as shown below:

```
Sleep(5000)
MouseClick("left",500,500,1)
Sleep(5000)
MouseClick("left",10,10,1)
```

Code sample 3.1 Script introducing Sleep.

Run the script, and you will see that the mouse moves to the same points as before, but with some delay, which was determined by Sleep. The unit of the delay is millisecond (ms, a thousandth of a second), so in our script the delay was 5 s each time.

At this point, you might be realizing that scripting can be very useful. Imagine that the first mouse click in our script is the "Ok" button at a software interface that controls an instrument, and the second click the analogous button to another instrument. If instead of a delay of 5 s you used a longer delay of, for example, 15 min, which is nearer the time span of many chemical analyses, you could be controlling two different instruments doing some type of measurement in concert. You might also be thinking that so far we could only set two mouse clicks, and that you would need to rewrite the lines of the script 50 times if you needed to measure 50 samples. You will see that this is not necessary. Before that, let us become familiar with other important and useful features of AutoIt.

3.2 Moving and Activating Windows

In order to be fully useful, mouse click control must ensure that every time the mouse click is activated it reaches the correct button or element on the software that is being controlled. This will not happen if the window of the program being controlled is at a different position than that when the script was written. The best way to ensure that a window will be at a given position is by using the function "WinMove." Let us see an example. This time, we will use the programs that were introduced in Chapter 1, FakeCarbonAnalyzerController (FACACO) and FAKe AutoSampler (FAKAS).

Open FACACO and FAKAS windows, and type the code below on SciTE:

```
opt("WinTitleMatchMode",1)
WinMove("FAke","",0, 0)
WinMove("FAKAS","",10, 10)
```

Code sample 3.2 Script introducing opt for windows title and WinMove function.

Make sure the SciTE window is not maximized (that is, not taking the full screen) and that you can see both FACACO and FAKAS windows on the screen. Save the script and press F5 to see what happens. You should have seen both windows moving to the top left corner of the screen, FAKAS a little more centered than FACACO (if you do not see it, it is because the FAKAS window is behind the FACACO window). Now let us examine the code. The "opt" instruction has as arguments "WinTitleMatchMode" and "1". opt is not properly a function, but an instruction that determines how some functions in a code will perform. If opt receives WinTitleMatchMode as an argument, it will determine how windows titles will be understood by functions dealing with windows, as WinMove, for example. If the next argument passed to opt is 1, it means that only the initial parts of the title of the window need to be used as an argument. For example, for FACACO instead of its long full title, "Fake" was used instead. If you do not call opt, the standard naming convention is used, which demands the full title to be passed as an argument. For the function WinMove, the first argument is the title of the window, the second the text, (which was left blank as it is optional), and the remaining two are the X and Y coordinates of the left upper corner of each window. In all cases in this book, the text of a window will not be passed as an argument for functions dealing with windows.

A problem in our previous code is that we could not determine which window would be in the foreground or background. Suppose that we want to ensure that FAKAS is on the foreground. This is the time to use the "WinActivate" function. See the code below:

```
opt("WinTitleMatchMode",1)
WinMove("FAke","",0, 0)
WinMove("FAKAS","",10, 10)
WinActivate("FAKAS")
```

Code sample 3.3 Script introducing WinActivate.

Run the script now and see what happens. You should see the FAKAS window in the foreground.

3.3 Sending Keyboard Inputs

Another very important function in AutoIt is "Send." MouseClick, which you are already familiar with, controls the mouse. Send, by its turn, controls the keyboard. For example, in FAKAS, we need to use the keyboard to indicate the position of the autosampler. Let us write a script that fills the input section of FAKAS to illustrate the Send function:

```
opt("WinTitleMatchMode",1)
WinActivate("FAKAS")
Send("6")
```

Code sample 3.4 Script introducing Send.

Save and run the script, and the number 6 should be typed inside the input area. Note, however, that the number that was there before remained there. For example, if the number 1 was there, the input area contents became 16, and not 6, after the script execution. However, we wanted the contents to become 6, and not 16. In order to resolve this, let us use the following script:

```
opt("WinTitleMatchMode",1)
WinActivate("FAKAS")
Send("{BACKSPACE 50}","{DEL 50}")
Send("6")
```

Code sample 3.5 Another script featuring Send.

Save and run the script, and now the number 6 should appear alone in the input area. The difference between this and the previous script is the line "Send("{BACKSPACE 50}","{DEL 50}")." In this line, we are saying that the backspace key should be typed 50 times and the delete key another 50 times. This way, we ensure that the input area will be empty unless a long input was in there before the script.

Send is a very powerful command that permits full control of the keyboard. The AutoIt help file entry for Send is a very useful reference.

3.4 "For" Loops and Variables

Controlling mouse clicks, positioning windows, and sending keyboard inputs as described so far are very powerful features of AutoIt. However, another aspect of AutoIt that makes it useful is the use of structured programming features, like loops and conditionals. This is where programming logic enters, and what can be a more difficult part for readers without a background in programming, because the way instructions are given is not the usual one that people use in their daily communication. However, once mastered, this skill proves invaluable. It is like riding a bicycle: you learn it once and never forget.

Let us start with a very simple example. Suppose that we want to repeat 10 times the mouse clicks in our code. As described in Section 3.1, you could simply copy and paste the code already written and press F5. However, there is another way of doing the same task, that is, by using a loop. A loop, in programming, is a block of code containing repeated instructions. A common type of loop is the "For … Next." Below is the code for 10 repetitions of our script using the "For … Next" loop:

```
For $step = 1 to 10
    sleep(5000)
    MouseClick("left", 500, 500, 1)
    sleep(5000)
    MouseClick("left", 10, 10, 1)
Next
```

Code sample 3.6 Script introducing "For … Next" loops.

If you type this in your script and run it by pressing F5, you will see the mouse moving 20 times with 5 s between each movement. This is a much more organized way of stating the instructions than repeating the original code (Code sample 3.1) 10 times. Let us have a deeper look at the Code sample 3.6 to ensure that you understand it.

First, it is necessary to understand that this is a block of code, and that the first line, "For $step = 1 to 10," only makes sense if there is a line with "Next" after it. In other words, in plain language, "For $step starting at 1 and going until 10, do what is written between here and Next." When Next is reached, "$step" is incremented by one unit. "For" is the instruction here. "$step" is a variable. This is the first time we meet a variable, but variables are very common in programming, and you will use them very often. Variables can receive values by means of the signal "=". In this case, the variable $step received 1, 2, 3, …, 10 each step of the loop. Note that the lines between For and Next are indented. This is a way of organizing the code, and makes clear that these lines belong to that part of the code. This is useful especially for long codes. SciTE automatically indents lines after instructions like "For".

If "For" loops are still confusing, with the following code you will probably understand them better. Before that, save your previous file in order to not lose the older code, and start a new one for the following:

```
For $step = 1 to 10
    MsgBox(0,"step",$step)
Next
```

Code sample 3.7 Another "For … Next" example.

Here, the MouseClick and Sleep functions are replaced for a single "MsgBox" function. MsgBox opens a small window and waits for you to press "Ok" to close it. If you press F5, this will happen 10 times, each time showing a window named "step" and with a text that is a number, which is, in fact, the value for the variable $step at each step of the loop. Note that when you pass a variable as an argument it does not go between " "; this is necessary only when passing literal words as arguments. If you are still confused, it is recommend to use the instructions presented so far until you understand them. Warning: it is possible to make infinite loops, and, just in case you make one by mistake, you can still stop the script by going to SciTE and pressing Ctrl + Break. The next sections also feature other examples of "For" loops that may help with their understanding.

3.4.1 Automating FAKAS

Let us discuss one more example using the "For ... Next" structure. This time, we will automate FAKAS. When introducing the "Send" command, we wrote a script using FAKAS. Let us modify it a little so that it will do the same action several times:

```
opt("WinTitleMatchMode",1)
WinActivate("FAKAS")
For $sample = 1 to 10
    Send("{BACKSPACE 50}","{DEL 50}")
    Send($sample)
    Sleep(1000)
Next
```

Code sample 3.8 Use of "For ... Next" loop for FAKAS.

Run the script, and you will see the value in the input field changing every second from 1 to 10. Let us now imagine that FAKAS is really controlling an autosampler, and that we want to see it moving from sample 1 to 10. This will demand that clicks are done in the buttons of the FAKAS interface, and will combine everything we learned so far, as shown in the script below:

```
opt("WinTitleMatchMode",1)
WinMove("FAKAS","",0,300)
For $sample = 1 to 10
    WinActivate("FAKAS")
    Mouseclick("left",80,360)
    Send("{BACKSPACE 50}","{DEL 50}")
    Send($sample)
    Sleep(500)
    WinActivate("FAKAS")
    Mouseclick("left",80,380)
    Sleep(10000)
    WinActivate("FAKAS")
    Mouseclick("left",80,410)
    Sleep(10000)
    WinActivate("FAKAS")
    Mouseclick("left",80,440)
    Sleep(10000)
Next
```

Code sample 3.9 Script demonstrating the automation of FAKAS.

Before running this script, be aware that it will take a while to be finished, and that the mouse will keep moving to the determined positions, meaning that it will be difficult to perform other activities with the computer while the script is running. If you want to stop it in the middle of the action, go to the SciTE window, open the Tools menu, and select Stop.

Let us have a closer look at the code. The first two lines were explained in Section 3.2, so have a look there to understand them. The loop, as we learned here, states that the actions between For and Next will be repeated 10 times (the variable $sample changing from 1 to 10). The actions are a sequence of window activations, mouse clicks, and sleeping periods. They are organized so that first the position of the sample is input in the input area, then the sample is moved to that position by the click of the button "Go to sample," then the needle is lowered to the sample, and finally the needle is returned to the up position. The sleeping times ensure that the buttons will be active at each click, and the window activations ensure that the mouse clicks will be done on the buttons, and not in another window.

You might be asking how I determined the sleeping times. In most cases, it is possible to measure the time that the actions take to happen and then use a sleeping time slightly longer than that. For positioning mouse clicks, trial and error is an option, but better ways are to use the AutoIt v3 Windows Info utility, or the AU3Recorder, which will be presented in the next sections.

3.4.2 First view of AutoIt v3 Windows Info (AWI)

AutoIt v3 Windows Info (abbreviated as AWI) is a utility that helps considerably when writing scripts. It was designed to provide the necessary information that needs to be supplied to functions in scripts. Here, only the information regarding mouse clicks will be discussed. Other features of AWI will be presented as they become relevant through the text.

You should find the executable file for AWI together with the one for AutoIt. Alternatively, it can be accessed from the Tools menu on SciTE provided the script you are writing was saved (the content of the menus change depending on this). By starting AWI, you should see a window that is always on top of the others. The window looks like Figure 3.1.

There are several tabs in the AWI window. For our purposes now, the most useful one will be the "Mouse" tab. Select this tab, as shown in Figure 3.1. The most important feature of AWI is the finder tool. Left-click on the "target" below the "Finder tool" label and then, holding down the button, move the mouse somewhere else on the screen, and then finally release the button. You should see information regarding the position, the cursor id (technical information that does not matter now), and a code for the color of the pixel, where the mouse pointer was released. Therefore, using this technique it is possible to get the appropriate position for the mouse clicks in scripts.

Now let us see another tab on AWI that can be useful for positioning windows. Choose the window tab on AWI, and it should look like Figure 3.2.

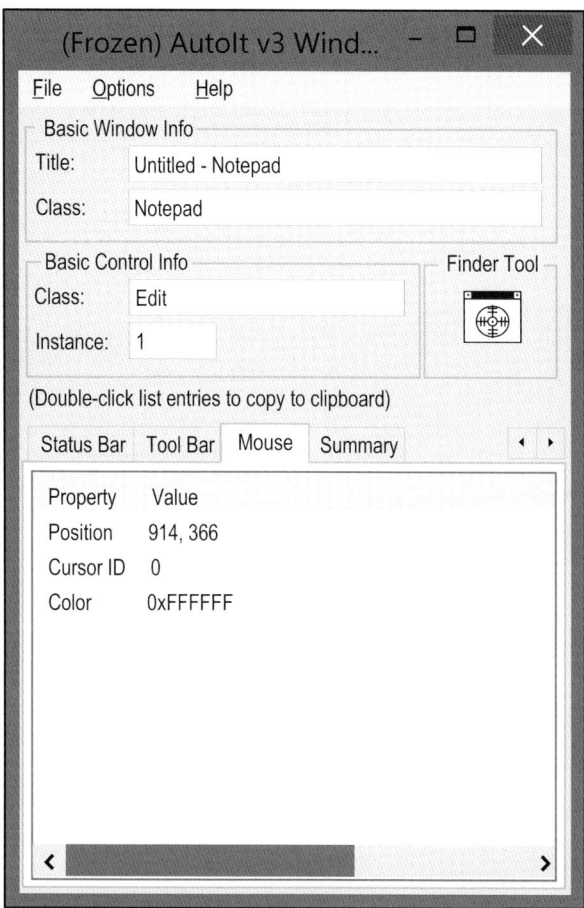

Figure 3.1 AutoIt v3 Windows Info (AWI) window showing mouse properties.

Figure 3.2 Window tab on AWI for FAKAS.

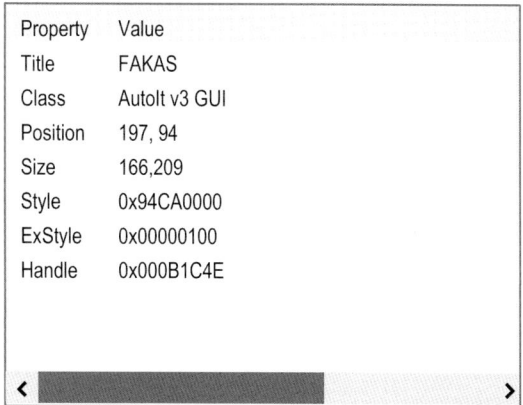

Note that you can get the window position from AWI. This can be useful if you need a window to be at a specific position and you need to know its coordinates. Most of the remaining information is not normally necessary for scripting. The title is the other important information on display, but you already know that you can get it from the window itself, without needing AWI, as we have been doing so far. The advantage of using AWI, although, is that you can copy the title and do not need to type it, thus avoiding misspelling.

3.4.3 AU3Recorder

In addition to AWI, AutoIt also has a recorder of mouse and keyboard inputs. This recorder makes it easy to track the position of mouse clicks. In order to use it, go to SciTE. Make sure you have a saved script open (it can be an empty one, but it needs to be saved). In the Tools menu, you will find AU3Recorder listed. Click on it, and a small window will appear. You can click on the area indicated "Click to Record" (Figure 3.3) to start recording your actions.

When you click on the Click to Record area, the window will disappear and instead another small window saying "Click to Stop" will appear. While this window exists, your actions keep being recorded, only stopping when you click on it. When stopping the recording, the recorder window closes and immediately some lines of code are input in SciTE (e.g., see Code Sample 3.10). The automatically generated code is much more complex than what we are doing now, so let us ignore most of it. Just concentrate on the MouseClick function calls for now. Look that they have the coordinates already input for each of them. Also, see that each Send function is called with already the proper arguments inside. Finally, there is a function called _WinWaitActivate. Its most useful feature is that the function automatically gets the full title name of the window that has been clicked.

```
+>SciTEDir => C:\Program Files (x86)\AutoIt3\SciTE UserDir => C:\
SCITE_USERHOME => C:\
#region --- Au3Recorder generated code Start (v3.3.9.5
KeyboardLayout=00000409)    ---
_WinWaitActivate("Untitled - Notepad","")
MouseClick("left",158,9,1)
Send("haosihaosid{ENTER}hoasias{ENTER}jhoasioas{ENTER}")
MouseClick("left",160,11,1)
#endregion --- Au3Recorder generated code End ---
```

Code sample 3.10 Example of code automatically generated by Au3Recorder.

Overall, AU3Recorder does a good job for getting screen coordinates and other useful information. It can also be used to generate simple scripts, but I personally do not like its approach, as the code generated is more complex than it needs to be in most cases. Nevertheless, AU3Recorder can be a useful tool after you get used to it.

Figure 3.3 Area indicated as "Click to Record" in The AU3Recorder utility.

3.4.4 Automating FACACO

If you tested FACACO while reading Chapter 2, you may recall how annoying was to keep answering the warning window every time it showed up. So, let us write a script that automates the clicking on that window:

```
Opt("WinTitleMatchMode", 1)
WinMove("FAke", "", 0, 0)
WinActivate("FAke")
MouseClick("left", 50, 40)
Sleep(500)
WinActivate("Start run")
Send("{ENTER}")
Sleep(3 * 1000)
For $sample = 1 To 7
   WinActivate("Warning")
   Send("{ENTER}")
   Sleep(50 * 1000)
Next
```

Code sample 3.11 Script demonstrating the automation of FACACO.

Make sure that seven sample names are typed in the Sample column and that the Analysis column for each sample contains dissolved inorganic carbon (DIC) or dissolved organic carbon (DOC), and that the Status for these samples is "Waiting." Run the script, and you should see the FACACO window moving to the top left corner of the screen, followed by the mouse clicking on the Start button, then by the pop-up window opening and closing, and then the next pop-up window opening and closing too, and finally by the seven samples being "measured," as indicated by the Status column and the opening and closing of pop-up windows. As for the example for FAKAS, the script consists mostly of windows activations, mouse clicks, and sleep times.

It is fundamental to understand the "For … Next" structure well in order to write useful scripts. It will be used in many examples throughout this book. Please review the contents of this section for a better understanding.

3.5 Organizing Your Code: Functions and Libraries

In the previous section, you learned how to automate FACACO and FAKAS separately. You may be thinking that now it is only a small step to make them work together, and you are right. You could try it yourself by combining the two scripts for each program in a single one, paying attention to the sequence of actions. However, this way you would write a very long sequence of lines that would be hard to modify if you needed to change window positioning or timing. Here, we will combine the codes but in a much more organized way, which is by writing our own functions.

As you learned in Section 2.4, functions are blocks of code that perform a well-defined task. So far, we have dealt with several functions like MouseClick, Sleep, Send, WinActivate, and WinMove. In addition to providing a large number of functions, AutoIt allows you to write your own functions. In our case, we are going to write some functions that for FACACO or for FAKAS and combine them in a code that synchronizes the two programs:

```
opt("WinTitleMatchMode",1)
WinMove("FAKAS","",0,300)
WinMove("FAke","",0,0)
FACACOstart()
For $sample = 1 to 3
   FAKASGoTo($sample)
   FAKASNeedleDown()
   FACACOmeasure()
   FAKASNeddleUp()
Next
Func FAKASGoTo($input)
   WinActivate("FAKAS")
   Mouseclick("left",80,360)
   Send("{BACKSPACE 50}","{DEL 50}")
   Send($input)
   Sleep(500)
   WinActivate("FAKAS")
   Mouseclick("left",80,380)
   Sleep(6000)
EndFunc
Func FAKASNeedleDown()
   WinActivate("FAKAS")
   Mouseclick("left",80,410)
   Sleep(5000)
EndFunc
Func FAKASNeddleUp()
   WinActivate("FAKAS")
   Mouseclick("left",80,440)
   Sleep(5000)
EndFunc
Func FACACOstart()
   WinActivate("FAke")
   MouseClick("left",50,40)
   Sleep(500)
   WinActivate("Start run")
   Send("{ENTER}")
   Sleep(3*1000)
EndFunc
Func FACACOmeasure()
   WinActivate("Warning")
   Send("{ENTER}")
   Sleep(50*1000)
EndFunc
```

Code sample 3.12 Script demonstrating the synchronous automation of FACACO and FAKAS.

Before running the script, let us analyze the Code sample 3.12; it is simpler than it may look from its relatively large size. The code is separated in "paragraphs." The first paragraph deals with determining the way of dealing with window title and positioning the windows. The second one is the main part of the code, where the different functions are called. Below this paragraph are the five functions defined for this code. Note that the first and fourth functions take arguments (in this case, the variable $input), while the other three functions do not need arguments. The remaining of the code is similar to those in previous examples and should be easily understood by the reader.

Note that the main part of the code (fourth line and For loop following it) is very clear to understand: First start FACACO, then repeat three times the next moves: move the needle to the sample, then put it down, do the analysis, and finally bring the needle back up. These five actions are described in detail in each function. Each function, by its turn, performs a well-defined action that consists of several smaller actions.

Still before running this script, make sure that you have three samples listed in FACACO, as previously explained, including the type of analysis (DIC or DOC), and that the needle position on FAKAS is up. By running the script, you will see that the two programs will work in synchrony, enabling, in this case, the simulated automation of two instruments in the correct way so that three samples would be measured. This may be more clearly seen in Figure 3.4.

3.5.1 Calling Functions from Different Files

The functions written in Code sample 3.12 can be reused in other scripts. The obvious way is to simply copy and paste them at the end of each code as done in the example. However, this makes scripts unnecessarily long, and also creates the problem that if we modify a function in one script we need to modify it on each different script featuring it.

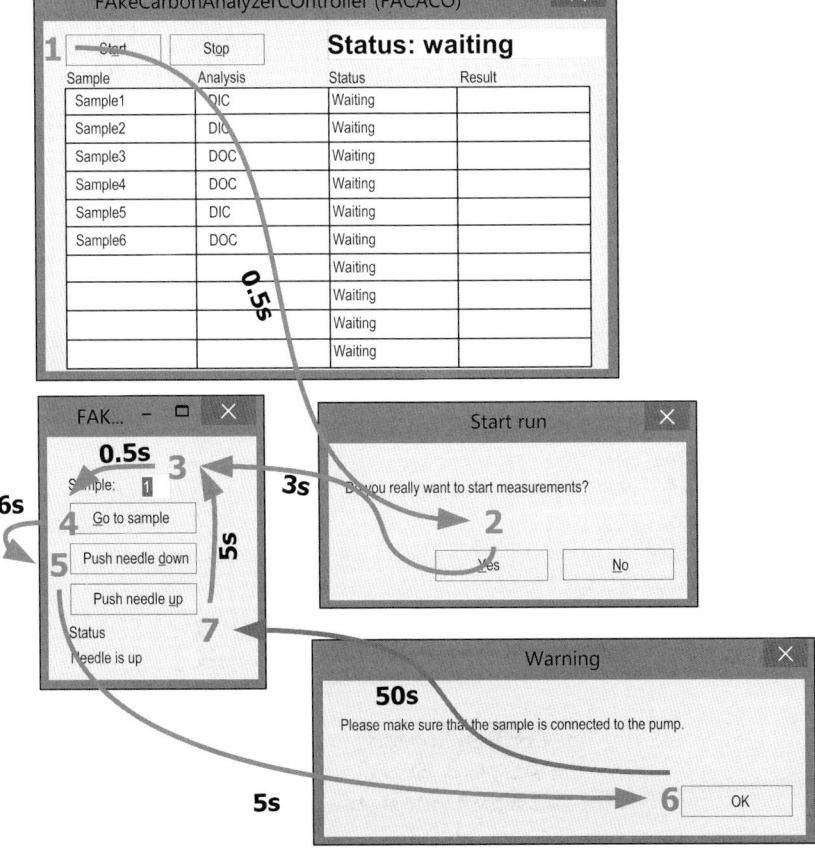

Figure 3.4 Flow of action of the script in Code sample 3.12. Larger numbers are the steps at which mouse clicks or keyboard inputs are done. Arrows link the steps. Smaller numbers near each arrow show the time in seconds between each action linked by the arrow.

A much better solution is to save the functions in a separated file and call them whenever necessary. The only requisite is that this file is in one of the folders that AutoIt checks when running a script, an easy pick being the same folder as the file of the main code. For our case, simply copy the five functions in Code sample 3.12 to an empty file on SciTE and save the file as "FACACOFAKASfunctions.au3." Then, write the following script:

```
#include "FACACOFAKASfunctions.au3"
opt("WinTitleMatchMode",1)
WinMove("FAKAS","",0,300)
WinMove("FAke","",0,0)
FACACOstart()
For $sample = 1 to 3
    FAKASGoTo($sample)
    FAKASNeedleDown()
    FACACOmeasure()
    FAKASNeddleUp()
Next
```

Code sample 3.13 Script introducing the #include command thus enabling the access of user-defined functions in the file FACACOFAKASfunctions.au3.

By using the keyword #include, it is possible to access the functions that were defined already in the other file. In the AutoIt community, libraries are often referred as *user-defined functions* (UDFs). Many UDFs are available for download from the AutoIt website, especially from the user forum.

3.6 Replacing Mouse Clicks with Keyboard Shortcuts

If a script needs to be implemented on another machine using another monitor, scripts relying on mouse click need to be tested to ensure that the clicks will be done at the right positions. Using keyboard shortcuts, this concern is not necessary.

AutoIt can automate keyboard shortcuts instead of mouse clicks when those shortcuts are available for the software. FACACO and FAKAS enable keyboard shortcuts, which can be seen by the underlined letters in the buttons controlling the actions in the software. For example, go to FAKAS and press CTRL + g; this is equivalent to pressing the "Go to sample" button. What we need to do now is to write the functions that replace mouse clicks with keyboard shortcuts and add them to the FACACOFAKAS-functions.au3 file. Here are the functions:

```
Func FAKASGoToShortcut($input)
    WinActivate("FAKAS")
    Mouseclick("left",80,360)
    Send("{BACKSPACE 50}","{DEL 50}")
    Send($input)
    Sleep(500)
    WinActivate("FAKAS")
    Send("^g")
    Sleep(6000)
EndFunc
```

3.6 Replacing Mouse Clicks with Keyboard Shortcuts

```
Func FAKASNeedleDownShortcut()
   WinActivate("FAKAS")
   Send("^d")
    Sleep(5000)
EndFunc
Func FAKASNeddleUpShortcut()
   WinActivate("FAKAS")
   Send("^u")
    Sleep(5000)
EndFunc
Func FACACOstartShortcut()
   WinActivate("FAke")
   Send("^a")
   Sleep(500)
   WinActivate("Start run")
   Send("{ENTER}")
   Sleep(3*1000)
EndFunc
```

Code sample 3.14 Modified functions to control FACACO and FAKAS using numerous keyboard shortcuts instead of mouse clicks.

Note that the functions are very similar to those in Code sample 3.12, but most mouse clicks were substituted by keyboard shortcuts. Also note that there are Sleep calls after the shortcut commands. It is always a good idea to leave some time between keyboard shortcuts so that, if any unexpected delay happens, the script will not be affected by it.

As you can see in the script, the way AutoIt called keyboard shortcuts was using ^ together with g, d, or u. This command sent the Ctrl key together with one of the letters. It is also possible to control the Alt (!) and Shift (+) keys in the same way, which enables the use of virtually any keyboard shortcut available in different programs.

It was not possible to eliminate the mouse click that goes inside the input area in FAKAS, because there is no keyboard shortcut for that action. Therefore, implementing keyboard shortcuts is not always a complete replacement for mouse clicks and window positioning.

Remember to save the FACACOFAKASfunctions.au3 file containing the new functions. It is possible now to write a script calling them, as shown below:

```
#include "FACACOFAKASfunctions.au3"
opt("WinTitleMatchMode",1)
WinMove("FAKAS","",0,300)
WinMove("FAke","",0,0)
FACACOstartShortcut()
For $sample = 1 to 3
   FAKASGoToShortcut($sample)
   FAKASNeedleDownShortcut()
   FACACOmeasure()
   FAKASNeddleUpShortcut()
Next
```

Code sample 3.15 Synchronous automation of FACACO and FAKAS using the functions using keyboard shortcuts (see Code sample 3.14).

Run the script, always paying attention first to the condition FACACO is (number of samples, etc.). Its behavior must be the same of the script in Code sample 3.13.

3.7 Summary

- This chapter presented the simplest way of synchronizing different programs, which is using timed mouse clicks or keyboard entries.
- Sleep is a function that sets a time delay.
- WinMove is a function that moves windows to a determined position.
- WinActivate is a function that activates a window, that is, makes it accessible to mouse clicks and keyboard entries.
- Send is a function that sends keyboard entries.
- For … Next loops allow the definition of repetitive actions for a determined number of steps.
- Variables are elements of the code that can receive values. In AutoIt, they are indicated by $.
- MsgBox is a function that opens a small window.
- AWI is a very useful tool that allows, among many other things, to easily find mouse click coordinates on the screen.
- AU3Recorder is another utility that can be used to find mouse click coordinates on the screen.
- You can organize your code using functions and libraries.
- Functions are blocks of code that perform one or more tasks.
- Libraries are collections of functions.
- It is possible to reuse libraries in new codes by calling them using #include.
- It is better to rely on keyboard shortcuts than on mouse clicks, when possible.

4

Interactive Scripting

In Chapter 3, it was demonstrated that even a very basic knowledge of AutoIt is enough to enable useful automation of procedures in a laboratory. That approach, however, had a limitation: the scripts relied on timing for all actions. This can be a problem if the analysis being performed has an unpredictable duration. For example, titrations can be short or long depending on the concentration of the substance to be measured in the sample. For such cases, setting a very long sleeping time in the script will ensure that almost all samples will be measured correctly. However, if this sleeping time is long, and most samples have lower analysis time, efficiency will be reduced. More sophisticated control is possible by the use of interactive scripting, that is, scripting that relies on not only inputs by the programmer but also outputs from the computer. In this chapter, we will learn how to write these "smarter" scripts.

The main form of computer output is messages on the screen. Signals like "ready," "finished," and "warning" are common in analytical software. Many times, these signals come as pop-up windows. Other times, the software controlling the instrument uses indicators that are not windows, but simple text messages or color changes in buttons. With AutoIt, it is possible to write scripts that wait for windows, or for an element on the screen to change its color. This chapter demonstrates the use of these techniques.

4.1 Window Monitoring

We saw in Chapter 3 that AutoIt can move and activate windows. AutoIt can also check if a given window is active, closed, or even exists at all. Write the following script:

```
WinWait("FAKAS")
MsgBox(0,"","FAKAS appeared!")
```

Code sample 4.1 Script introducing WinWait.

Run the script. Nothing happens unless you start FAKe AutoSampler (FAKAS). Then, as soon as it opens, a pop-up window appears, as expected from the script. Other similar functions are "WinWaitActive" and "WinWaitClose." WinWaitActive waits for a window to become active, that is, if you click on it, for example. WinWaitClose waits for a window to be closed. It is a good idea to test these functions by writing simple scripts.

We can improve our script that synchronizes FACACO and FAKAS using window monitoring. First, it is useful to rerun that script so that we can observe in detail what it does. You should already have written Code samples 3.14 and 3.15 and created a file named FACACOFAKASfunctions.au3 containing the functions called in Code sample 3.14. Remember: it is necessary that the functions are listed in the FACACOFAKASfunctions.au3 file, and that this file is in the same folder containing the script file. Furthermore, FACACO and FAKAS must be open, and there must be three samples listed in FACACO, including the analysis type (dissolved inorganic carbon (DIC) or dissolved organic carbon (DOC)), and that the status is "Waiting" for each of the three samples. Observe that after a sample is finished, the pop-up window with title "Warning" appears and the computer stays idle for some time until the needle is pulled up and the next sample starts. This unnecessary delay comes from the functions FACACOmeasure, which calls a delay of 50 s after the button is pressed, and from the function FAKASNeedleUp, which is called only after this 50-s delay (see the codes for these functions in Code sample 3.14). In order to optimize our script, we will define three new functions:

```
Func FAKASGoToWinWait($input)
   WinWait("Warning")
   WinActivate("FAKAS")
   Mouseclick("left",80,360)
   Send("{BACKSPACE 50}","{DEL 50}")
   Send($input)
   Sleep(500)
   WinActivate("FAKAS")
   Send("^g")
   Sleep(6000)
EndFunc
Func FACACOstartWinWait()
   WinActivate("FAke")
   Send("^a")
   WinWait("Start")
   WinActivate("Start run")
   Send("{ENTER}")
EndFunc
Func FACACOmeasureWinWait()
   WinActivate("Warning")
   Send("{ENTER}")
   Sleep(10*1000)
EndFunc
```

Code sample 4.2 New functions for synchronizing FACACO and FAKAS using the WinWait function.

Include the new functions in the file "FACACOFAKASfunctions.au3" and save it. Then, rewrite the code that calls the functions like this:

```
#include "FACACOFAKASfunctions.au3"
opt("WinTitleMatchMode",1)
WinMove("FAKAS","",0,300)
WinMove("FAke","",0,0)
FACACOstartWinWait()
For $sample = 1 to 3
   FAKASGoToWinWait($sample)
   FAKASNeedleDownShortcut()
   FACACOmeasureWinWait()
   FAKASNeddleUpShortcut()
Next
```

Code sample 4.3 Script for synchronization of FACACO and FAKAS using window monitoring.

You can try this script instead of the original one (Code sample 3.15) and see what happens. Always remember to input sample names, analysis type, and measuring status in FACACO before running the script. When you run the script, you will see two big differences in relation to the original one: (i) the needle is pulled up before the status of the sample being measured is "Done," but after it is "Sampling." In real situations, this would be Ok, since after sampling there is no need for the needle to be in the vial; this was achieved by reducing the Sleep time in FACACOmeasureWinWait to 10 s (Code sample 4.2), instead of 50 s in FACACOmeasure (Code sample 3.11) and (ii) samples are started as soon as the Warning pop-up window is closed, which is a direct effect of using WinWait; in the previous code, the subsequent samples would start some seconds after this window was closed. There were other less noticeable differences between the two scripts. You may have noticed that a difference between FAKASGoToWinWait (Code sample 4.2) with FAKASGoToShortcut (Code sample 3.14) is the WinWait command on the first line of FAKASGoToWinWait. There is no such line in the previous function. The reason for including this line in the newer function is that in the previous script for synchronization between FAKAS and FACACO it did not matter where to locate the sleep command. However, in the new synchronization based on window monitoring, it was necessary that the waiting step was localized in the function dealing with the Go To button. Another difference between the scripts is between the functions FACACOstartWinWait (Code sample 4.2) and FACACOstart (Code sample 3.12). The short sleep of 0.5 s was replaced by a WinWait command, which may have saved a fraction of second each run. The differences between the current script based on window monitoring and the one based on timing can be better understood by comparing the action flows of the two scripts (Figures 3.3 and 4.1).

Overall, the new script is faster than the original one. In the example given here, this amounts to only a few seconds per sample, and our example has only three samples. In real situation, when tens or hundreds of samples can be measured unattended, and each analysis can take at least 5 min, any gain in the range of 1 min is very significant.

4.2 Pixel Monitoring

Window monitoring is very useful for making interactive scripts. However, it is not every time that pop-up windows are used as outputs by the software controlling the equipment. Sometimes, only a message like "done," "measuring," or "ready" is shown; this message is often shown using different colors so that it becomes more visible. With AutoIt, it is possible to take advantage of these signals and write interactive scripts.

Let us start with a very simple script. First, let us choose a pixel to deal with. The easiest way to do that is using the AutoIt v3 Windows Info (AWI) utility. In Chapter 3, AWI was used for finding the coordinates of a mouse click. You may remember that, when doing that, the color code of the pixel was also shown. It is also possible to obtain the pixel color using the function "PixelGetColor." See the following example:

```
$Pcolor = PixelGetColor(500,500)
MsgBox(0,"",$Pcolor)
MsgBox(0,"",Dec("FFFFFF"))
```

Code sample 4.4 Introducing PixelGetColor.

4 Interactive Scripting

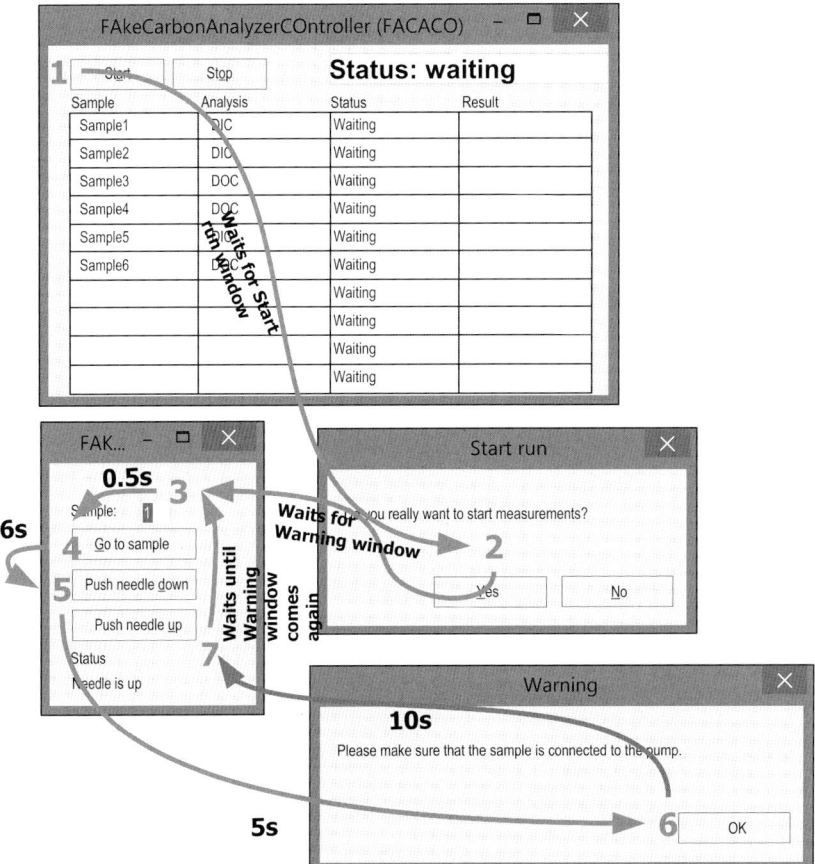

Figure 4.1 Flow of action of the script in Code sample 4.3. Large numbers are the steps at which mouse clicks or keyboard inputs are done. Arrows link the steps. Smaller numbers near each arrow show the time in seconds between each action linked by the arrow. The text in bold black between steps explains that the subsequent step only comes after the action described there takes place.

Run the script. Two message boxes will appear, each with a different number. The number will be 16777215 in the second window. The number on the first depends on what is being sampled at the 500,500 pixel. Try to ensure that this pixel is white by, for example, opening a Notepad window and positioning it so that the writing area covers the pixel at position 500,500. You can verify this by using AWI and checking that the color code for the pixel is FFFFFF. Then run the script. Both pop-up windows now should show 16777215, which is the decimal equivalent of FFFFFF, the white color code in hexadecimal. Note that in the second MsgBox, command in the script the function "Dec" is used to convert from hexadecimal to decimal. In summary, remember that the value obtained from AWI is the same as that obtained using PixelGetColor, but they are at different numerical bases, and need to be converted in order to be comparable.

AWI has a "magnifying glass" that can be useful to find the right pixel when writing scripts based on pixel monitoring. On the AWI window, go to the Options menu, and

Figure 4.2 AWI magnifying glass showing pixels in detail for a section of the screen containing plain black text on a white background.

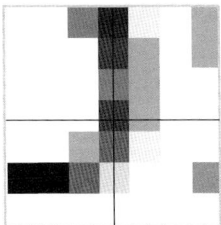

then select magnifying glass. When you move the Finder Tool again, you will see that a square with a close-up of the neighborhood of the mouse pointer will appear (Figure 4.2).

Note how elaborate and unexpected can be the pixel colors: in Figure 4.2, the mouse pointer (not shown as when printscreen is used for a screen shot it disappears) was near one of the letters typed in Notepad. You could think that it would be all black and white, but in fact there are several colors making up each character. This illustrates how the magnifier can be useful when getting the right color of a pixel.

4.3 "While … WEnd" Loops for Pixel Monitoring

Make sure that the 500,500 position shows a pixel that is NOT white. Now write the following script:

```
$Wcolor = 16777215
$Pcolor = PixelGetColor(500,500)
While $Wcolor <> $Pcolor
       sleep(20)
       $Pcolor = PixelGetColor(500,500)
WEnd
MsgBox(0,"","Script finished!")
```

Code sample 4.5 Introducing "While … WEnd" loops.

Run the script. Nothing will happen unless you make the pixel 500,500 white again; as soon as you do it, a window pops up announcing that the script is finished. The script is setting two variables: $Wcolor and $Pcolor. In AutoIt, you set the value of a variable using the "=" operator, and use the same sign to state equality. In this script, in all cases, the equal sign is being used to attribute a value to a variable. In the "While … WEnd" loop, the value of $Pcolor changes every 20 ms. At the moment it stops being different (operator <> in the script) from $Wcolor, the loop stops and the pop-up window appears. If you understood the "For … Next" loop presented in the previous chapter, the "While … WEnd" loop should not be difficult to grasp. The fundamental difference between the two structures is that in the "For … Next" loop, a variable has its value automatically changed in the "For" line itself. By contrast, the key variable in a "While … WEnd" loop is not changed unless explicitly stated in the script. In Code sample 4.5, the first two lines state that the variable $Wcolor and $Pcolor will certainly be different. Then, in the loop, variable $Pcolor is checked every 20 ms until it is not different from

$Wcolor anymore. Writing this code using a "For ... Next" loop would be complex, because it would be necessary to set up a maximum value for the key variable, and in principle the loop in Code sample 4.5 could last forever. Let us see another example of "While ... WEnd" loop in order to make it clearer:

```
$value = 0
While $value <1000
    $value = InputBox("Checking numbers","Write a number:")
    Sleep(20)
WEnd
```

Code sample 4.6 Another example of a "While ... WEnd" loop.

Run the script. You will see a pop-up window asking for a number (this pop-up window was created using the function "InputBox"). You can type a number and press the Ok button. It will keep coming back unless you type 1000 or more in the writing field. In the first line of the code, the variable $value receives the value of 0. This is to ensure that when the While loop starts, it will be lower than 1000, because, as you can see, it is stated that while $value is less than 1000, what is inside the loop must be done. The While instruction is usually followed by an expression involving one of the following operators: > (larger than), < (less than), = (equal to; the same character used for attribution, see Section 3.4), and <> (different). Inside the loop, there are two lines of code: one line in which $value receives the value of the InputBox function, and the next is a sleep of 20 ms.

4.4 Synchronizing FACACO and KAKAS Using Pixel Monitoring

Suppose that the warning pop-up did not appear when using. In this case, it would be necessary to use pixel monitoring to achieve interactive scripting. After using FACACO several times for doing exercises, you may have noted that the background of the word "Status" sometimes became green. If you paid enough attention, you noted that it got green every time the status of a sample became "Done" in the spreadsheet. We will take advantage of this output of FACACO and write a script so that it is not necessary to rely on timing only (like that in Code sample 3.15). As for when we wrote the modified script based on window monitoring (Code samples 4.2 and 4.3), we will change both the main code and the functions in the file FACACOFAKASfunctions.au3.

The first step is to get the color code of one of the green pixels around the word status. First, we run FACACO for a single sample and wait until it ends. The window must be at position 0,0 (easily achieved by running one of the previous scripts so far in the book, e.g., Code sample 3.2). The background around Status will be green. Then, using the procedure like the one in Section 4.2, we find that the pixel at position 252,52 can be used, and that the color code for it is 65280 (Figure 4.3).

Using this information, we can adapt the code in Code sample 4.5 to synchronize FACACO and FAKAS, creating a new function, FACACOpixel:

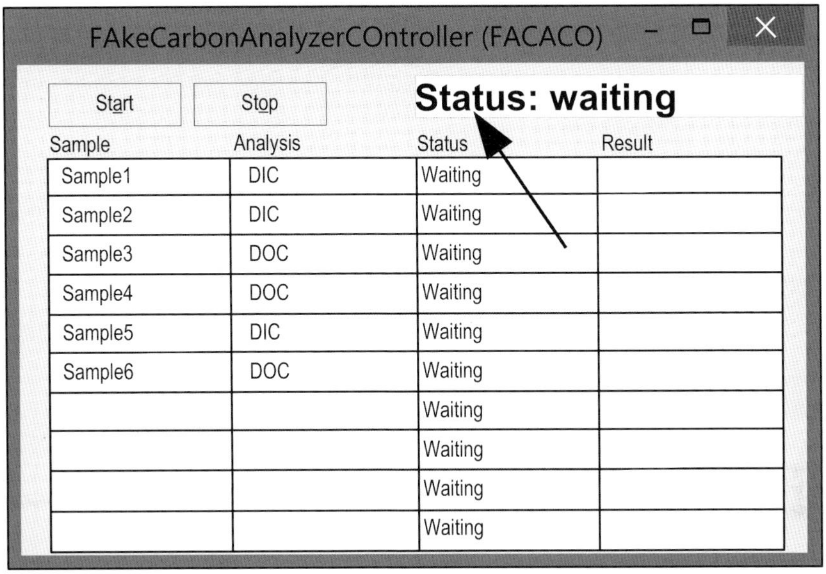

4.4 Synchronizing FACACO and KAKAS Using Pixel Monitoring

Figure 4.3 FACACO at position 0,0 on the screen, and the tip of the arrow at position 252,52.

```
Func FACACOpixel()
    $Wcolor = 65280
    $Pcolor = PixelGetColor(252,52)
    While $Wcolor <> $Pcolor
        sleep(20)
        $Pcolor = PixelGetColor(252,52)
    WEnd
EndFunc
```

Code sample 4.7 Adaptation of pixel watching for FACACO.

Include this function in the FACACOFAKASfunctions.au3, save it, and then write the following script:

```
#include "FACACOFAKASfunctions.au3"
opt("WinTitleMatchMode",1)
WinMove("FAKAS","",0,300)
WinMove("FAke","",0,0)
FACACOstartShortcut()
$sample = 1
For $sample = 1 to 3
    Sleep(3*1000)
    FAKASGoToShortcut($sample)
    FAKASNeedleDownShortcut()
    FACACOmeasureWinWait()
    FAKASNeddleUpShortcut()
    FACACOpixel()
Next
```

Code sample 4.8 FACACO and FAKAS synchronized using pixel monitoring.

42 | *4 Interactive Scripting*

The script in Code sample 4.8 has no new elements, and you should be able to understand it by comparing it with Code sample 4.3. Remember that the function FACACOmeasureWinWait does not have a WinWait command (Code sample 4.2). Thus, it relies only on timing and pixel monitoring. Running the script, you will see that it has efficiency comparable to the window monitoring-based script. However, scripts based on window monitoring are preferable to scripts based on pixel monitoring, because it is necessary that the pixel being watched be always at the correct position, while a window being monitored can be at any position in the screen.

The action flow of the script based on pixel monitoring (Figure 4.4) can be compared to the previous ones (Figures 3.3 and 4.1) for better understanding.

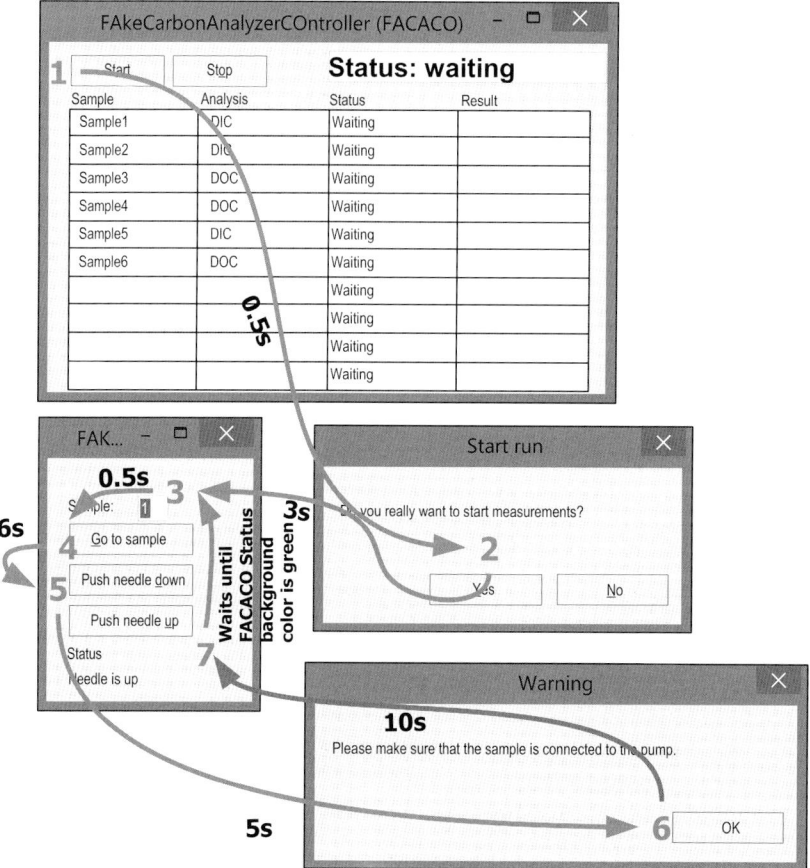

Figure 4.4 Flow of action of the script in Code sample 4.8. Large numbers are the steps at which mouse clicks or keyboard shortcuts are done. Arrows link the steps. Smaller numbers near each arrow show the time in seconds between each action linked by the arrow. The text in black between steps 7 and 3 explains that the subsequent step only comes after the action described there takes place.

4.5 Enhanced Pixel Monitoring Using PixelCheckSum

In addition to monitoring the color of a single pixel on the screen, AutoIt can also monitor a group of pixels by using the function "PixelCheckSum." This function can be more useful than PixelGetColor in some cases. For example, suppose we want to monitor the status field on FAKAS. This status field changes its contents according to the actions being performed by FAKAS. However, the background color does not change and thus it is difficult to select a single pixel to be monitored, which would be a pixel that in a given status would be gray and, in another, black, because it would be occupied by a character of the status message. Furthermore, if the selection is not done carefully, it is possible that the same pixel gets black for two different statuses, which would make the script unreliable. Using PixelChekSum, the sum of the color codes for all pixels in a selected area is used, which virtually ensures that a different number will be obtained for each different status message. Let us see an example:

```
#include "FACACOFAKASfunctions.au3"
opt("WinTitleMatchMode",1)
WinMove("FAKAS","",0,300)
WinMove("Untitled","",500,300)
WinActivate("FAKAS")
Send("^g")
MsgBox(0,"Waiting for the right time","Close me at the right time")
$sum = PixelChecksum(10,480,150,500)
WinActivate("Untitled")
Send($sum & "{ENTER}")
```

Code sample 4.9 Script that allows getting the value for PixelCheckSum for different actions done using FAKAS.

FAKAS and Notepad both need to be open in order that the script works. Run the script. If FAKAS had the needle in the up position (as it should at the start), the status message should have changed to "Moving to sample," then "Arriving at sample," and finally "Needle is up." While these messages were shown on the FAKAS window, a pop-up window appeared asking you to close it at the right time. By doing so, it gets the sum of the pixel values for the square between pixels 10,480 and 150,500 and types it to Notepad. The value should be 1549482090 if the message was "Moving to sample," 668813236 if it was "Arrived at sample," and "657711511" if "Needle is up." Now, let us substitute "^g" for "^d" in Code sample 4.8, and run again the script. If you close the pop-up window when the message is "Needle is down," you should get the value 2973926643. In addition to PixelChecSum, Code sample 4.9 introduces the operator "&" (used as an argument for Send), which is a concatenator, that is, it connects two chains of characters. This operator is very useful for the Send function, because it allows mixing variables (like $sum, in our code) and characters. In this chapter, AWI was used to determine that the pixels 10,480 and 150,500 were useful to encompass the area of interest.

The purpose of Code sample 4.9 was to provide the necessary values to use the PixelCheckSum function when synchronizing FACACO and FAKAS. We can now use them in functions with synchronization in mind:

```
Func FAKASGoToPixelSum($input)
    WinActivate("FAKAS")
    Mouseclick("left",80,360)
    Send("{BACKSPACE 50}","{DEL 50}")
    Send($input)
    Sleep(500)
    WinActivate("FAKAS")
    Send("^g")
    $Psum = PixelChecksum(10,480,150,500)
    While $Psum <> 668813236
        sleep(20)
        $Psum = PixelChecksum(10,480,150,500)
    WEnd
    Sleep(2000)
EndFunc
Func FAKASNeedleDownPixelSum()
    WinActivate("FAKAS")
    Send("^d")
    $Psum = PixelChecksum(10,480,150,500)
    While $Psum <> 2973926643
        sleep(20)
        $Psum = PixelChecksum(10,480,150,500)
    WEnd
EndFunc
```

Code sample 4.10 Functions that use PixelCheckSum for the status area of FAKAS and can be used to synchronize FAKAS and FACACO.

Include these functions in FACACOFAKASfunctions.au3. These functions are modifications of those listed in Code sample 3.14. If you look carefully, you will see that the main modification was to replace some Sleep commands by a While loop that uses PixelCheckSum for each different status shown on FAKAS. The strategy was similar to that used in Code sample 4.7. Now, write the following script:

```
#include "FACACOFAKASfunctions.au3"
opt("WinTitleMatchMode",1)
WinMove("FAKAS","",0,300)
WinMove("FAke","",0,0)
FACACOstartShortcut()
For $sample = 1 to 3
    Sleep(3*1000)
    FAKASGoToPixelSum($sample)
    FAKASNeedleDownPixelSum()
    FACACOmeasureWinWait()
    FAKASNeddleUpShortcut()
    FACACOpixel()
Next
```

Code sample 4.11 FACACO and FAKAS synchronized using enhanced pixel monitoring.

Take the usual precautions for FACACO and FAKAS (see explanation after Code sample 3.11) and run the script. Compared with the previous version (Code sample 4.7), the

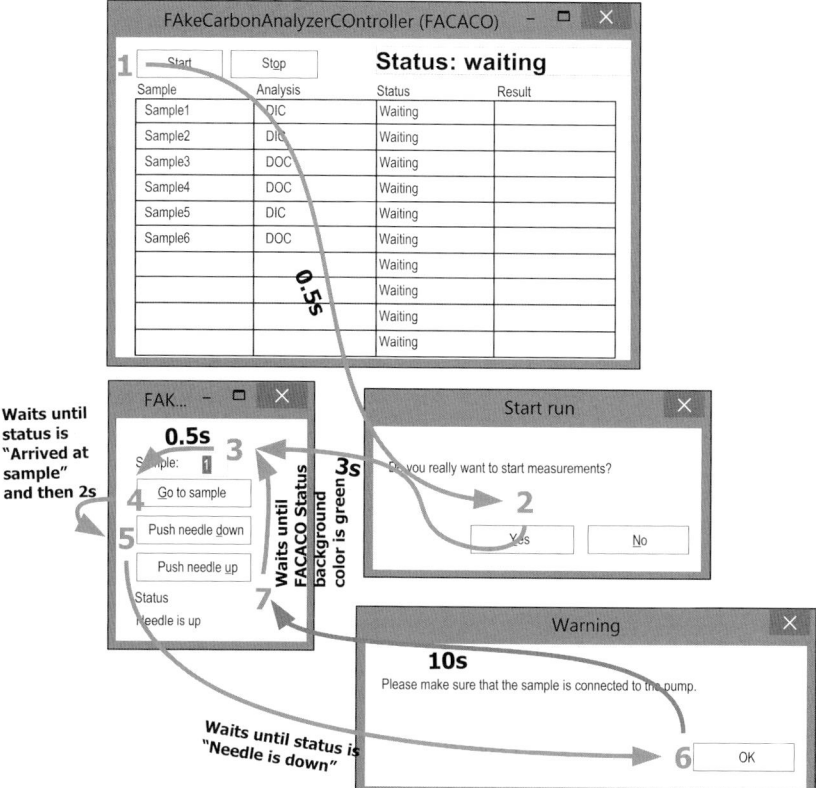

Figure 4.5 Flow of action of the script in Code sample 4.11. Large numbers are the steps at which mouse clicks or keyboard inputs are done. Arrows link the steps. Smaller numbers near each arrow show the time in seconds between each action linked by the arrow. Texts in black between pairs of steps explain that the subsequent step only comes after the actions described there take place.

script is a little faster, because there are less arbitrary waiting times. However, a greater advantage here is that if a problem occurred with the fictional autosampler controlled by FAKAS, and this problem was reflected on a lack of change in the status field, the script would stall, which would be good, avoiding a potential loss of samples.

The flow of action for Code sample 4.11 is shown in Figure 4.5. It can be compared with the previous ones for better understanding.

You may be thinking that it would be possible to improve even more the code in order to eliminate other arbitrary waiting times, like that between steps 6 and 7 by employing PixelCheckSum for those fields in FACACO showing the status of a given measurement. This is true, and you can do it as an exercise, if you wish.

As a final note, it can be useful to know that even more sophisticated pixel analysis is possible. Optical character recognition (OCR) is the technology used to identify text in images. It is widely used by software processing scanned images, for example. AutoIt has libraries to deal with OCR that have the aim of identifying text on the screen. However, it is more complex than the procedures presented here, and does not work in all cases. More information about OCR for AutoIt is presented in Appendix B.

4.6 Blocking Access to Keyboard and Mouse

When using scripts based on mouse clicks, or when relying on pixel monitoring, it is fundamental that windows do not change their position and that they are in the foreground at the time of clicking. If there is concern that other users could access the computer while the script is running, and thus "ruining" the run, it can be a good idea to block access to the computer. AutoIt can ensure that even if someone gets access to the computer, he/she will not be able to control the mouse or keyboard while the script is running. The function that disables the keyboard and the mouse is "BlockInput":

```
#RequireAdmin
opt("WinTitleMatchMode",1)
BlockInput(1)
Sleep(5000)
MouseClick("left",500,500)
```

Code sample 4.12 Blocking keyboard and mouse control using Blockinput.

Run the script and, differently from all scripts presented so far, in this one a message from the Windows operating system comes asking if we want to proceed with the script. You should say yes, and will see that you cannot control the mouse or the keyboard while the script is running. Be very careful with this function, because if you write an infinite loop you could have your computer stuck! Fortunately, it is possible to unlock the computer using Ctrl + Alt + Del in the middle of the script.

4.7 Summary

- Despite being very simple, timed scripts are not efficient.
- Interactive scripts can save significant time and be more reliable than timed scripts.
- WinWait is a function that stops the script until a certain window appears.
- Pixel monitoring is a very useful technique when scripting.
- PixelGetColor is a function that gets a numeric value corresponding to the color of a pixel.
- AWI also can be used to get the color of a pixel. However, its value is returned in a different numerical base of that used by PixelGetColor. It is possible to convert the values using the function Dec.
- While … WEnd loops allow the setup of repetitive actions for an undetermined number of steps.
- While … WEnd loops are used with mathematical operators, which can be = (equal), <> (different), < (smaller than), and > (larger than), among others.
- Window monitoring is preferable to pixel monitoring, mainly because for pixel monitoring a window needs to be at a precisely fixed position, which is not necessary for window monitoring.

- PixelCheckSum is a function that gets a numeric value corresponding to the sum of the color values of a given area on the screen. It can be useful to identify messages on the screen.
- In order that they work properly, scripts presented in this and the previous chapter can benefit of restricted computer access, which can be set using the function BlockInput.

5

Scripting with Controls

In addition to keyboard shortcuts, there is another way through which AutoIt can control software without mouse clicks, that is, by using "controls." Controls, also known as *widgets*, are the elements found in graphical user interfaces (GUIs), like buttons, scroll bars, labels, and status bar. Therefore, all Windows users are familiar with controls. However, using AutoIt it is possible to access controls without the mouse or keyboard, which is not possible for human users (touch screen is equivalent to mouse click, so it does not count).

In order to write scripts based on controls, it is necessary to know how to access them. AutoIt has a utility built solely for this purpose, the AutoIt v3 Windows Info (AWI), which you already used in the previous chapters to locate mouse click coordinates (Section 3.4.2) and pixel color codes (Section 4.2). Here, a much more in-depth operation of AWI is presented.

Before starting, you must be aware that scripting with controls will not work with many programs. Therefore, although potentially very useful, and arguably more elegant, scripting with controls is, in fact, less powerful than the simpler techniques presented in Chapters 3 and 4.

5.1 Using AWI to Get Control Information

The best way to find information about controls is using AWI. Open it, as explained in Section 3.4.2, and this time leave the tab on Control, as shown in Figure 5.1.

As in the previous examples featuring AWI, some technical information is on display. Of all that, the most important for us is "ClassnameNN," which is the "name" of the control in the script code. As you may remember, you obtain this information by using the finder tool on AWI and moving it to the Go to sample button on FAKe AutoSampler (FAKAS). By moving the finder tool to the input field where you can type the sample number, the information obtained is different from that for the button (Figure 5.1). In addition to ClassnameNN, an important information for input fields is "Text," which in our example has a value of "1." However, sometimes AWI will fail to display the text of a control. Fortunately, you can also obtain this information by scripting, as shown in Section 5.2.

5 Scripting with Controls

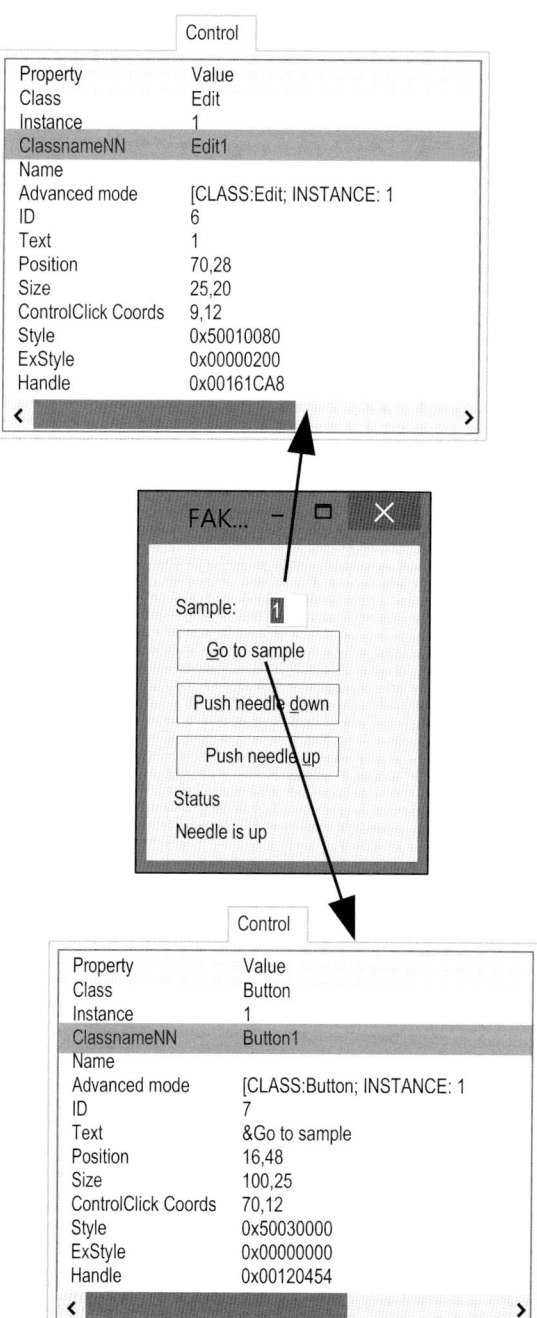

Figure 5.1 AutoIt v3 Windows Info (AWI) showing information about the "Go to sample" button and input field on FAKAS.

5.2 Functions That Provide Control Information

In order to write scripts using controls, it is essential that relevant information about them can be obtained during script execution. This is done using functions designed for this purpose. For example, see the code below, which demonstrates how using the function "ControlGetText" it is possible to obtain the same information as obtained using AWI:

```
opt("WinTitleMatchMode",1)
$word = ControlGetText("FAKAS","","Static3")
Msgbox(0,"",$word)
```

Code sample 5.1 Demonstration of ControlGetText.

Run the script. You will probably see the message "Needle is up" in the pop-up window. This is because the control Static3 refers to the Status at the bottom of the window. Try clicking on one of the buttons and quickly running the script again (or write another script to do that instead of you!). You should see different messages according to what the status becomes.

We can try the same approach with FAke Carbon Analyzer Controller (FACACO). Open FACACO and, using AWI, get the information about the cell just below the Status label. It should contain the word "Waiting." You will see that the cell is identified as Edit21, but that text is empty. However, if you go to the "Visible Text" tab on AWI, you will see all the "Waiting" instances there, together with other texts of other controls. Although AWI does not show "Waiting" in the text specific for this control, it is possible to get this information using the ControlGetText function, as shown in the script below:

```
opt("WinTitleMatchMode",1)
$word = ControlGetText("FAk","","Edit21")
Msgbox(0,"",$word)
```

Code sample 5.2 Using ControlGetText with FACACO.

Another function that is useful to get the status of a control is "ControlCommand." See the code below:

```
opt("WinTitleMatchMode",1)
$ButtonStatus = ControlCommand("FAKAS","","Button1","IsEnabled")
MsgBox(0,"Status",$ButtonStatus)
```

Code sample 5.3 Demonstration of ControlCommand.

Make sure FAKAS is open, and run the script. The message box should show "1" when you run the script. Now, click on the Go to sample button on FAKAS and quickly run the script again. If you are fast enough, the message box will show 0. You may have noted that when the button is enabled for being clicked, the function ControlCommand returns 1, and when the button is not enabled, it returns 0.

Knowing the status of a control is like knowing that a pop-up window came or that a pixel changed its color: it is a type of output. Therefore, it is amenable to approaches analogous to those presented in Chapter 4. This will be further explored in Section 5.4.

5.3 Sending Commands to Controls

The other necessary attribute to use controls in a script is the ability to send them commands. Here, we will see some functions that can be used for this purpose. For example, see the code below:

```
opt("WinTitleMatchMode",1)
ControlSetText("FAKAS","","Edit1","8")
ControlClick("FAKAS","","Button1")
```

Code sample 5.4 Introducing ControlSetText and ControlClick.

Run the script. If you did not watch the FAKAS window, you may not even have noted that anything happened because it was all in the background. If you are in doubt, bring FAKAS window to the foreground and run the script again. You should see the Sample field changing to 8 and the button being pressed, but without the mouse doing it. Both "ControlSetText" and "ControlClick" have similar arguments: the title of the window, text (optional), control to be used, and, for ControlSetText, the input value. You may remember that in Section 3.6 it was not possible to eliminate the mouse click that activated the input field. Now this was achieved, and truly mouse-free scripting became possible. The demonstration will be given in the next section.

Finally, there is the function "ControlSend," which in principle does exactly the same that ControlSetText does. However, there are cases in which ControlSend works better than ControlSetText, and others in which the opposite is true. You will learn through experience, in this case.

5.4 Synchronizing FACACO and FAKAS Using Controls

The functions presented so far can be used to automate FACACO and FAKAS. As when we learned about keyboard shortcuts, we first will write new functions combining the ones presented in this chapter:

```
Func FAKASGoToControl($input)
   ControlSetText("FAKAS","","Edit1",$input)
   ControlClick("FAKAS","","Button1")
   $ButtonStatus = ControlCommand("FAKAS","","Button1","IsEnabled")
   While $ButtonStatus=0
      Sleep(20)
      $ButtonStatus =
ControlCommand("FAKAS","","Button1","IsEnabled")
   WEnd
EndFunc
Func FAKASNeedleDownControl()
   ControlClick("FAKAS","","Button2")
   $ButtonStatus = ControlCommand("FAKAS","","Button2","IsEnabled")
   While $ButtonStatus=0
      Sleep(20)
      $ButtonStatus =
ControlCommand("FAKAS","","Button2","IsEnabled")
   WEnd
EndFunc
```

5.4 Synchronizing FACACO and FAKAS Using Controls

```
Func FAKASNeddleUpControl()
   ControlClick("FAKAS","","Button3")
    $ButtonStatus = ControlCommand("FAKAS","","Button3","IsEnabled")
   While $ButtonStatus=0
       Sleep(20)
       $ButtonStatus =
ControlCommand("FAKAS","","Button3","IsEnabled")
   WEnd
EndFunc
Func FACACOstartControl()
   ControlClick("FAk","","Button1")
   WinWait("Start")
   ControlClick("Start","","Button1")
EndFunc
Func FACACOmeasureControl($sample)
   ControlClick("Warning","","Button1")
   While ControlGetText("FAk","","Edit"&20+$sample) = "Sampling"
       Sleep(20)
   WEnd
EndFunc
Func FACACOstatusControl()
   $target = "Status: Done"
   $word = ControlGetText("FAk","","Static5")
   While $word <> $target
       sleep(20)
       $word = ControlGetText("FAk","","Static5")
   WEnd
EndFunc
```

Code sample 5.5 Functions based on controls to synchronize FACACO and FAKAS.

As with all functions that we created so far, save the new ones in the FACACOFAKAS-functions.AU3 file. The functions in Code sample 5.5 are very different from the equivalent ones shown in the previous chapters. They do not rely on fixed timing between actions; instead, there are While loops that do the timing, which is always dependent on the status of a given control. Note that on the third line of the function FACACOmeasureControl, we pass as argument referent to the control for the command ControlGetText "Edit"&20 + $sample. If you remember, & is a concatenator (introduced in Code sample 4.9), which is used to combine chains of characters. In our case, it connects the word "Edit" with the result of the operation "20 + $sample," which needs to be a number between 1 and 10 in order to work, otherwise other columns will be monitored. It works because on the Status column the topmost cell is named Edit21, the one below it Edit22, and so on until the bottom one, which is Edit30 (Figure 5.2).

The functions in Code sample 5.5 can be utilized in a script to synchronize FACACO and FAKAS:

```
#include "FACACOFAKASfunctions.au3"
opt("WinTitleMatchMode",1)
FACACOstartControl()
For $sample = 1 to 3
   FAKASGoToControl($sample)
   FAKASNeedleDownControl()
   FACACOmeasureControl($sample)
   FAKASNeddleUpControl()
   FACACOstatusControl()
Next
```

Code sample 5.6 Script that synchronizes FACACO and FAKAS using controls.

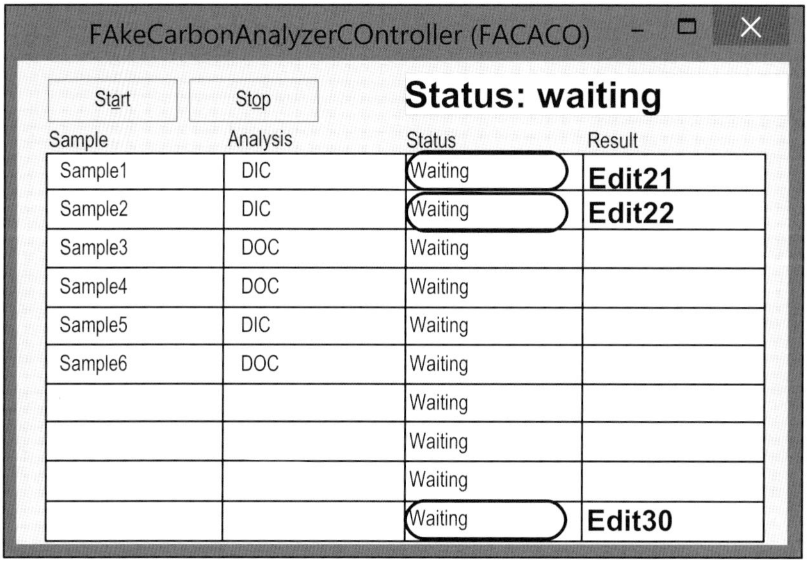

5 Scripting with Controls

Figure 5.2 Control names for some cells in FACACO.

Test the script with the usual precautions (see explanation for Code sample 3.12), and see how it performs. Note that actions will happen even without the windows being active. Compare the action flow for the script based on controls (Figure 5.3) with the previous ones in Chapters 3 and 4, and see that there are no fixed timings at all when using controls, this being the most striking difference.

A big advantage of using controls is that multitask is less affected by scripting than using the previous approaches. You can test the script in Code sample 5.4 again, but this time try doing something else at the same time, like surfing the web or writing a text on a word processor. The script should work seamlessly, even if FAKAS and FACACO windows are minimized. Because of that, it can be argued that scripting based on controls is the best way of achieving perfect synchronization between different software interfaces. However, as explained in the beginning of the chapter, it is not always possible to access the control information in some software. In these cases, mouse clicks or keyboard shortcuts are necessary. Also, despite the best efforts of the AutoIt team, there are cases in which a correctly written script using controls will fail to make them respond. The reasons for such failures are not clear. In all cases, it is essential that you test the script a few times to make sure that controls will respond to commands every time they are called. In my personal experience, controls are reliable in 99% of cases, and I use them daily in my work in the laboratory. A common problem is that the value in ClassnameNN changes sometimes after restarting the program. Therefore, it is advisable to check this parameter before running the script.

An alternative to the standard control functions available for AutoIt is the library for accessing UIA (user interface automation), which recognizes controls for some applications that are invisible for AWI, as, for example, those written in LabView, a

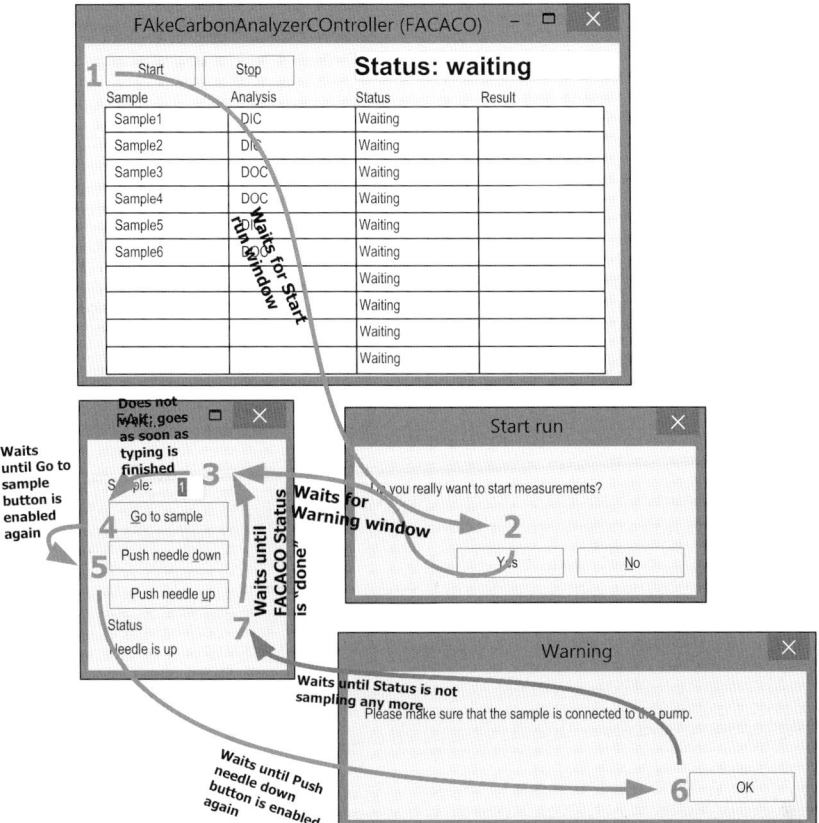

Figure 5.3 Flow of action of the script in Code sample 5.6. Large numbers are the steps at which mouse clicks (in this case, not real clicks, but the use of the function ControlClick) or keyboard inputs (also not real keyboard inputs, but ControlSend or ControlSetText) are done. Arrows link the steps. Smaller numbers near each arrow show the time in seconds between each action linked by the arrow. The texts between steps explain that the subsequent step only comes after the action described there takes place.

popular language for laboratory instruments. This alternative approach is discussed in Appendix C.

5.5 Dealing with Errors: If … Then

All scripts studied so far outlined actions to be taken linearly, that is, with fixed start and end points in all cases. However, sometimes it is necessary to be able to take alternative courses of action. AutoIt allows this type of approach by means of conditionals, which are instructions that determine which course of action will be taken depending on which condition is presented. Conditionals are fundamental elements of structured programming languages.

5 Scripting with Controls

Before introducing a useful application, let us see a very simple script that introduces If ... Then:

```
$first = InputBox("First number","Please type a number:")
$second = InputBox("Second number","Please another number:")
If $first > $second Then
    msgbox(0,"",$first)
Else
    msgbox(0,"",$second)
EndIf
```

Code sample 5.7 Introducing If ... Then.

Run the script. Two windows will show up, each asking for a number. After that, a third window comes showing the largest number. The two first lines in the code call the two first windows. The third line contains the If ... Then statement. It works like this: If the expression in the statement is true, then the next line is executed. If untrue, it goes to Else and does what is below it. Then, it finishes at EndIf. In this case, the script will work in such a way that always the larger of two numbers will be displayed after the questions.

An important use for conditionals is to deal with errors. A main source of errors is unexpected user behavior. For example, it is possible that, when using FAKAS, the user press the Go To Sample button before making the needle go up. While a pop-up window alerts the user for the problem, the scripts presented so far have no way to deal with that, and will continue their actions regardless of this problem. This is obviously not good, but AutoIt can deal with this by means of the structure If ... Then. See the following code:

```
#include "FACACOFAKASfunctions.au3"
opt("WinTitleMatchMode",1)
If ControlGetText("FAKAS","","Static3") = "Needle is up" Then
    FACACOstartControl()
    For $sample = 1 to 3
        FAKASGoToControl($sample)
        FAKASNeedleDownControl()
        FACACOmeasureControl($sample)
        FAKASNeddleUpControl()
        FACACOstatusControl()
    Next
Else
    MsgBox(0,"Error","Make sure needle is up on FAKAS")
EndIf
```

Code sample 5.8 Using If ... Then to avoid an error when synchronizing FACACO and FAKAS.

Run the script, following the usual precautions, but this time, before starting, press the Push needle down button on FAKAS before starting. You will see that a message asking you to bring the needle up will be shown, and the script will not run. If you bring the needle up, by pressing the Pull needle up button in FAKAS, and run the script again, it will run as it should. So, as in Code sample 6.5, the actions below the line containing If ... Then are only performed if the statement in the line is true. If not, the actions following Else are done.

If ... Then is a structure classified as a conditional. Conditionals are widely used in programming, and usually become more important as the codes become more complex.

5.6 Infinite Loops and Controls

Short scripts as those shown so far did not make use of conditionals, but if you need more stable scripts, that is, scripts that can deal with unexpected events, conditionals are essential. Also, as will be seen in Section 5.6, conditionals are useful for other types of procedures.

5.6 Infinite Loops and Controls

So far, the number of samples being measured was a key element in the scripts synchronizing two instruments (the number of steps in the For … Next loops). However, there are cases in which the number of samples is not important when synchronizing the instruments. For example, in the case of dissolved organic carbon (DOC) analysis, in some analytical setups a cryotrap is used to accumulate the CO_2 generated in the chemical reaction and release it as a concentrated pulse that enhances its measurement. The cryotrap can be a very simple device, like, for example, a linear actuator holding part of the gas lines over a liquid nitrogen Dewar. Suppose that we want to synchronize such a cryotrap that can move the gas lines down or up, thus immersing or removing them from the liquid nitrogen. Before showing the script that synchronizes this hypothetical cryotrap to FACACO, which is our hypothetical DOC analyzer, we need an interface to control the cryotrap. The code below generates this interface:

```
#include <GUIConstantsEx.au3>
$cryoValcoControl = GUICreate("cryo", 150,100,500,500)
$button1 = GUICtrlCreateButton("cryo Go",20,80,60,20)
$radio1 = GUICtrlCreateRadio("Up",20,20)
$radio2 = GUICtrlCreateRadio("Down",20,40)
GUISetState()
While 1
    Switch GUIGetMsg()
        Case $GUI_EVENT_CLOSE
            Exit
        Case $button1
    EndSwitch
WEnd
```

Code sample 5.9 Script that creates a graphical interface for a hypothetical cryotrap.

Run the script, and you should see a small window appearing (Figure 5.4). Do not try to understand Code sample 5.9 now; only after studying Chapter 18, you will be able to do that. For now, let us simply use the code to have our user interface to synchronize with FACACO. In order that we can run our script that we will write later, this script

Figure 5.4 Graphical user interface for a hypothetical cryotrap.

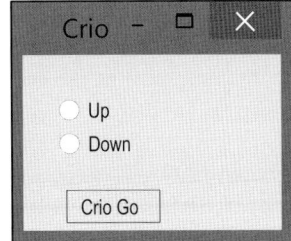

cannot be running at the same time. Thus, it is necessary that you compile it to a.exe file (by pressing Ctrl + F7 on SciTE).

Once you have your cryotrap controller working, make sure FACACO is open and has samples listed to be run. Make sure some of the samples to be run are dissolved inorganic carbon (DIC), and others are DOC. Now see the code below:

```
#include "FACACOFAKASfunctions.au3"
opt("WinTitleMatchMode",1)
While 1
    $FACACO_status = ControlGetText("FAk","","Static5")
    If $FACACO_status = "Status: Adding Oxidant" Then
        Sleep(100)
        CrioChangePosition("down")
        While $FACACO_status <> "Status: Measuring DOC"
            Sleep(100)
            $FACACO_status = ControlGetText("FAk","","Static5")
        WEnd
        CrioChangePosition("up")
    EndIf
WEnd
Func CrioChangePosition($position)
    If $position = "up" Then
        ControlClick("Crio","","Button2")
        Sleep(500)
        ControlClick("Crio","","Button1")
    ElseIf $position = "down" Then
        ControlClick("Crio","","Button3")
        Sleep(500)
        ControlClick("Crio","","Button1")
    EndIf
EndFunc
```

Code sample 5.10 Script that synchronizes FACACO and the cryotrap program generated in Code sample 5.9.

Now run the script. You should see that, for DIC samples, nothing happens to the cryotrap software. However, when you run DOC samples, you see that the radio buttons are activated alternately: the position down is activated immediately after the status of the sample being measured changes to Adding Oxidant, and the position up is activated when the status changes to Measuring . The script uses a function "CryoChangePosition," which is defined at the end of the code. The function is very simple, and has no new elements. The names of the controls were obtained using AWI. "Button1" corresponds to the "Cryo Go" button, "Button2" to the radio button labeled as "Up," and "Button3" to the radio button labeled "Down." Therefore, when the argument for the function is up, the radio button corresponding to Up is clicked, and then the Cryo Go button is clicked. The corresponding action to the radio button Down happens when down is passed as an argument.

The main part of the code consists of a While loop. The loop states "While 1," which means that the loop is infinite, and the only way to stop it is by closing the script. In every step of the loop, the variable "$FACACO_status" receives the value of the large label in FACACO (control Static5), which shows the status of the measurement. This happens every 100 ms, until this value is equal to "Status: Adding Oxidant," then the CryoChangePosition function is called with "down" as an argument, and another loop starts. This other loop lasts while $FACACO_status is different from "Status: Measuring DOC." Inside this loop, there is a pause of 100 ms and the variable $FACACO_status receives again the value of Static5 in FACACO. This means that every 100 ms this variable is checked. Once it becomes equal to "Status: Measuring DOC," the loop is finished and the outer loop resumes, starting the whole process over again. If you recall, this technique is very similar to that used for pixel watching presented in Chapter 4.

If you compile the script in Code sample 5.10 to a.exe file, it can work together with the script that synchronizes FACACO and FAKAS (Code sample 5.6). If you do so and observe the three programs, you will see that they all work in synchrony.

5.7 Summary

- The best way to synchronize different programs using AutoIt is by means of controls.
- Using controls, neither the mouse nor the keyboard needs to be accessed directly by the script, which allows the computer to be used for other purposes while the script runs.
- Controls are elements of program interfaces, such as buttons, labels, and menus.
- AWI is essential for working with controls.
- Not all programs have accessible controls, and thus it is not possible to use the technique for them.
- When using AWI to get information about a control, the most important parameters are ClassnameNN and Text.
- ControlGetText is a function that provides the text being displayed by a control.
- ControlCommand is another function that retrieves information from a control.
- ControlSetText is a function that changes the text of a control.
- ControlSend works very similarly to ControlSetText, but one is better than the other for specific contexts.
- Sometimes controls fail. Therefore, it is important that a script based on them be fully tested before put into action.
- An alternative approach to controls is presented in Appendix C. Using this alternative approach, some programs that are inaccessible to AWI can be automated. This is the case of many programs written in LabView, a popular language for software controlling laboratory instruments.
- If … Then is a type of conditional. It can be used to deal with errors that may occur during the execution of the script.
- Infinite loops can be used to synchronize programs when the processes being synchronized are not related to the number of samples being measured.

6

E-mail and Phone Alarms

Chapters 3–5 focused on the integration of (FAke Carbon Analyzer Controller) FACACO and FAKAS, which control ficticious laboratory instruments. However, AutoIt can synchronize such programs to any kind of software. In this chapter, we see how they can be synchronized to programs that allow communication through the Internet. The idea here is to demonstrate how simple alarms can be configured so that if something goes wrong in the middle of an unattended run, a message can be sent via e-mail, short message service (SMS), or even a phone call.

6.1 E-mail Alarms

There are two ways to automate e-mail sending with AutoIt: (i) using third-party software and (ii) directly using the appropriate network communication protocol.

6.1.1 Sending E-mail Using Third-Party Software

AutoIt can automate the sending of e-mails in many different ways. An obvious one is to use timed mouse clicks or keyboard keys while an e-mail interface is open. It is a simple modification of the techniques presented in Chapter 3, and has the advantages of simplicity and generality. First, we write a function:

```
Func SendGmailOpera($EmailAddress,$title,$message)
   opt("WinTitleMatchMode",1)
   WinActivate("Inbox")
   MouseClick("left",85,200)
   Sleep(1000)
   MouseClick("left",1200,320)
   Sleep(1000)
   Send($EmailAddress)
   Sleep(1000)
   MouseClick("left",1200,360)
   Sleep(1000)
```

Practical Laboratory Automation: Made Easy with AutoIt, First Edition. Matheus C. Carvalho.
© 2017 Wiley-VCH Verlag GmbH & Co. KGaA. Published 2017 by Wiley-VCH Verlag GmbH & Co. KGaA.

6 E-mail and Phone Alarms

```
    Send($title)
    Sleep(1000)
    MouseClick("left",1200,420)
    Sleep(1000)
    Send($message)
    Sleep(1000)
    MouseClick("left",870,700)
EndFunc
```

Code sample 6.1 Function that sends an e-mail using mouse clicks and keyboard entries to the Gmail interface on the Opera web browser.

The function "SendGmailOpera" receives as arguments the e-mail address that should receive the message, the message title, and the message itself. Then, it types them in the appropriate fields in the Gmail (www.gmail.com) interface when it is open in the Opera web browser. If you plan to use this code, you will probably need to adapt it to your e-mail interface, web browser (the window title may be different for other browsers), and monitor resolution. Mouse clicks coordinates can be determined more easily by following the approaches presented in Sections 3.4.2 and 3.4.3. Save the function to the file "Alarms.au3," and then write the following code:

```
#include "Alarms.au3"
$Sent = False
While $Sent = False
    If WinExists("Neddle error","") Then
        SendGmailOpera("myemail@lab.lab","Check FAKAS","Check FAKAS")
        $Sent = True
    EndIf
    Sleep(100)
WEnd
```

Code sample 6.2 Script that sends an automated e-mail after an error during FAKAS execution.

Run the script. You will probably see nothing happening, unless you open FAKAS, send the needle down, and then try to move the needle to a certain position. As you probably already know from practicing through the examples in the previous chapters, in this condition the autosampler cannot move and an error message will pop up warning us of that (Figure 6.1). This script will check if this error comes (using the function "WinExists" inside an If … Then statement, which is explained in Section 5.5), and, if so, starts a function ("SendGmailOpera") that sends the message (its title and contents are the same, see line 5) to a predefined e-mail account ("myemail@lab.lab"; you will need to change this in your code for a real e-mail account) and changes $Sent to True.

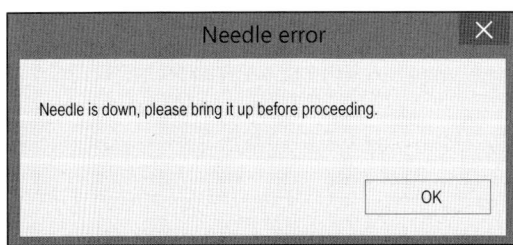

Figure 6.1 Error message window generated by FAKAS if the needle is down and you try to move it to a different position.

Because when $Sent is True it is no longer False, the loop stops. The Sleep at the end of the loop determines the pace of WinExists. A time value of 100 ms has been chosen here, but shorter or longer times can be used as per requirement.

Despite being very simple and easy, as we learned relying on mouse and keyboard inputs has the weakness of vulnerability to user action (which can be resolved by using BlockInput, see Section 4.6) and, in this case, of the chance that the e-mail web page will close down, or another web page will open on top of the one being used, and so on. A solution is to use an e-mail client that is independent of the web browser, such as Outlook and Thunderbird, to mention only two among the most popular. The use of such software will not be demonstrated, as from the viewpoint of AutoIt the procedure is nearly the same as that one in Code sample 6.1.

6.1.2 Sending E-mail Using SMTP

A more sophisticated approach to sending e-mails is by making direct use of the Simple Mail Transfer Protocol (SMTP). When you access your e-mail using an Internet browser or e-mail client, you are using SMTP, but this is not apparent for you. AutoIt allows you to send e-mails without needing a browser or e-mail client by means of directly using SMTP. I found a user-defined function (UDF) that works very well to send e-mails using Gmail on https://www.autoitscript.com/forum/topic/23860-smtp-mailer-that-supports-html-and-attachments/, developed by Jos. I modified it to make it simpler and more suited to the purposes of sending just a simple message:

```
1)  Func SendMailSimple($Server, $Sender, $Destination, $Subject, $Contents, $Username, $Password, $IPPort, $ssl)
2)      Local $objEmail = ObjCreate("CDO.Message")
3)      $objEmail.From = '"' & $Sender & '" <' & $Sender & '>'
4)      $objEmail.To = $Destination
5)      Local $i_Error = 0
6)      Local $i_Error_desciption = ""
7)      $objEmail.Subject = $Subject
8)      If StringInStr($Contents, "<") And StringInStr($Contents, ">") Then
9)          $objEmail.HTMLBody = $Contents
10)     Else
11)         $objEmail.Textbody = $Contents & @CRLF
12)     EndIf
13)     $objEmail.Configuration.Fields.Item ("http://schemas.microsoft.com/cdo/configuration/sendusing") = 2
14)     $objEmail.Configuration.Fields.Item ("http://schemas.microsoft.com/cdo/configuration/smtpserver") = $Server
15)     If Number($IPPort) = 0 then $IPPort = 25
16)     $objEmail.Configuration.Fields.Item ("http://schemas.microsoft.com/cdo/configuration/smtpserverport") = $IPPort
17)     If $Username <> "" Then
18)         $objEmail.Configuration.Fields.Item ("http://schemas.microsoft.com/cdo/configuration/smtpauthenticate") = 1
19)         $objEmail.Configuration.Fields.Item ("http://schemas.microsoft.com/cdo/configuration/sendusername") = $Username
20)         $objEmail.Configuration.Fields.Item ("http://schemas.microsoft.com/cdo/configuration/sendpassword") = $Password
```

```
21)     EndIf
22)     If $ssl Then
23)         $objEmail.Configuration.Fields.Item
("http://schemas.microsoft.com/cdo/configuration/smtpusessl") = True
24)     EndIf
25)     $objEmail.Configuration.Fields.Update
26)     $objEmail.Fields.Item ("urn:schemas:mailheader:Importance") =
"Normal"
27)     $objEmail.Fields.Update
28)     $objEmail.Send
29)     $objEmail=""
30) EndFunc
```

Code sample 6.3 Function that sends an e-mail using Gmail without a browser via SMTP.

Copy the script without the line numbers (e.g., "2)"). These line numbers were added to this code to show that some long lines should be typed in SciTE as a single line. Other codes throughout the book will have line numbers; always remove them if copying the code to the Alarms.au3 file. This code will not be explained, as it is more advanced than the others in this book. Interested readers can study the original code supplied in the forum. By the way, the forum available at www.autoitscript.com is a very useful learning environment, and it is a good idea to become a member of the community and participate in the forum in order to improve your skills on AutoIt.

A script making use of the function in Code sample 6.3 that sends an e-mail after the error of needle positioning in FAKAS can be like:

```
1) #include "Alarms.au3"
2) $Sent = False
3) While $Sent = False
4)     If WinExists("Neddle error","") Then
5) SendMailSimple("smtp.gmail.com","youemail@gmail.com","destination@a
ddress.net","Warning, error with FAKAS!","The needle was at the wrong
position, please fix this in order to continue
operation!","username","password",465,1)
6)         $Sent = True
7)     EndIf
8)     Sleep(100)
9) WEnd
```

Code sample 6.4 Automated e-mail sending using SMTP after an error during FAKAS execution

Make sure you replace "youemail@gmail.com," "destination@address.net," "username," and "password" with appropriate entities (first, you must have a Gmail account). Again, the line numbering must be removed when copying the code (note that line 5 is long). Also, and very importantly, you must disable a security setting on Gmail that disables the sending of e-mails other than by a web browser. You will see how to do that when you try to use this script the first time. Gmail will send you an e-mail explaining what you need to do (it will first advise you not to do it, but you can carry out the procedure without any problem). After all these steps, you can run the script and should see the e-mail being sent to the destination as expected if the error window (Figure 6.1) is generated by FAKAS. You may be curious about what "465" and "1" mean in the function call. Putting in the simplest possible way, they are necessary codes that make using Gmail possible. These codes can be different for other e-mail providers.

It is beyond the scope of this book to explore such details, and interested readers can study this subject further by reading the Internet forum about AutoIt, and resources on network protocols like TCP/IP and SMTP.

6.2 SMS and Phone Call Alarms

In some cases, it can be necessary to use other means of communication than e-mail. SMS and phone calls are widespread technologies, and thus alarms based on them have the potential to reach many useful destinations, such as older model cell phones and even landlines.

6.2.1 Sending SMS

Using short message service (SMS) and phone calls, you can reach ordinary cell phones that cannot access e-mail, and even landline phones, if you are giving phone calls. We will use Skype, one of the most popular software packages for computer-based messaging, video talk, and phone calls to give a simple warning in the case of a problem in the execution of an analytical procedure.

The first thing you need to do is to install Skype. You can download the installation files from http://www.skype.com. Once Skype is installed, you can set up an account and then communicate using the Internet for free. Also, if you connect a credit card or bank account to your Skype account, you can also send SMS and call landline phones. Therefore, these services are not free, but cost very little.

Sending SMS using Skype is very easy, as you will see. First, let us examine the Skype window. Open Skype and press Ctrl + D. You should see the input field as in Figure 6.2.

Note that if you type anything, it appear in that field. To the right of the field, there are two figures in gray circles, one with a telephone and the other with a text balloon. Send TAB a few times, and you will see that you can select either one of them. Select the text balloon and press enter. The writing focus will change to the bottom of the screen (Figure 6.3). For example, in Figure 6.3, "Warning!" is written in that field.

With these properties in mind, let us write a function that automates SMS sending via Skype:

```
Func SendSMSSkype($number, $message)
   WinActivate("Skype")
   Send("^d")
   Send($number)
   Sleep(1000)
   Send("{TAB 4}")
   Sleep(1000)
   Send("{ENTER}")
   Sleep(1000)
   Send($message)
   Sleep(1000)
   Send("{ENTER}")
   Send("{ENTER}")
EndFunc
```

Code sample 6.5 Function that sends automated SMS using Skype.

Figure 6.2 Input field for phone numbers on Skype window after pressing Ctrl + D.

Figure 6.3 New input field that appears after the text balloon (Figure 6.2) is chosen.

Save the code to Alarms.au3, and now you can write the code for dealing with FAKAS error:

```
#include "Alarms.au3"
$Sent = False
While $Sent = False
    If WinExists("Neddle error","") Then
        SendSMSSkype(044777,"Check FAKAS, it has a problem!")
        $Sent = True
    EndIf
    Sleep(100)
WEnd
```

Code sample 6.6 Automated SMS sending using Skype after an error during FAKAS execution.

Note that a fake phone number (044777) has been used. Of course, you need to use a real phone number in order that someone gets the message. Run the script, and you should see Skype doing everything automatically if the error message (Figure 6.1) exists. A good aspect of Code sample 6.6 is that it does not rely on mouse clicks, which means that the Skype window can be at any position. Be aware that this code worked for version 7.17.0.105. It is not warranted that it will work for other versions of the program. In the version used, the sequence of actions was: first, the Skype window opened with the WinActivate function; second, the input area for sending phone calls or SMS opened after sending Ctrl + D; third, the destination number was typed; fourth, the TAB key was sent four times, which meant that the focus went to the SMS icon on the Skype interface; fifth, the focus changed to the typing area after the ENTER key was sent, the message was then written; and finally, ENTER was sent twice (in this specific case, sending ENTER only once did not work).

6.2.2 Making Phone Calls

Making a phone call suits the extreme case in which you can only access a landline phone, and not a cell phone, and thus cannot receive SMS. As will be seen, this procedure is more complex than the others presented so far. The main reasons are that while an e-mail or SMS will almost surely reach their destiny when sent, phone calls only work if the destination answers the phone. Therefore, it is necessary to deal with potential problems in addition to simply sending the call.

Also, sending a phone call demands that an audio record exists in first place, differently from text messages, which can be defined in the script itself. Thus, the first step is to record a voice message. You can use any program that enables such feature, like for example "Sound Recorder", available for Windows 8.1. For this example, we will use a file named "WarningMessage.wma", full path: C:\WarningMessage.wma).

The next step is being able to play the recorded file. Again, there are many alternatives, the easiest one being to use the "SoundPlay" function available for AutoIt.

Finally, you need a microphone that can be moved near the speakers of your computer (I assume that you have a computer that has speakers, which emit sounds) so that the message played can be heard when the phone call is made.

We will use Skype to make the phone call. We will do it manually first so that the script will become easier to understand. The first step is to press Ctrl + D, so that the interface to make calls or SMS is activated (Figure 6.2). Then, we can type the phone number and press the call button (the phone symbol, Figure 6.2). Once this is done, the call starts. You can terminate the call by pressing Alt + PageDown. You can test and answer the call. Note that the only way to know if someone answered the call is by looking at the screen. In other words, there are no pop-up windows, text messages, or controls changing status. Therefore, pixel monitoring is necessary to know when a call is successful, and this implies the need of fixed window positioning (remember what you learned in Chapters 4 and 5). With these characteristics in mind, we can write the functions that will enable automated phone calls:

```
Func CallSkype($number)
   WinActivate("Skype")
   WinMove("Skype","",0,0,500,471)
   Send("^d")
   Send($number)
   Send("{TAB 3}")
   Send("{ENTER}")
   Sleep(5000)
EndFunc
Func TerminateCallSkype()
   WinActivate("Skype")
   Send("!{PGDN}")
   Sleep(2000)
EndFunc
Func RepeatUntilReachSkype($interval,$x1,$y1,$x2,$y2,$PixelSum)
   $reached = False
   $start = TimerInit()
   While $reached = False
      If TimerDiff($start) < $interval Then
         If PixelChecksum($x1, $y1, $x2, $y2) <> $PixelSum Then
            $reached = True
         EndIf
      Else
         TerminateCallSkype()
         CallSkype(0266269565)
         $start = TimerInit()
      EndIf
      Sleep(100)
   WEnd
EndFunc
```

Code sample 6.7 Functions to make a phone call using a voice message stored in the computer using Skype.

Figure 6.4 Dial figure on Skype window during a phone call.

Save the functions in Alarms.au3. The first function, "CallSkype," is very similar to SendSMSSkype (Code sample 6.5). The differences are the number of TAB clicks (because the phone icon, and not the message icon, is chosen, see Figure 6.2) and that the function WinMove is called. In addition to the coordinates (0,0), we are passing the new dimensions in pixels (500 in width and 471 in height) for the Skype window. This is necessary because, as explained above, once the phone call is answered, it is necessary to rely on pixel monitoring.

The following function, "TerminateCallSkype," is very simple, without any new element. This function activates the Skype window, sends Alt + PageDown, and waits for 2 s. By this way, it terminates a call.

The next function, "RepeatUntilReachSkype," is the most complicated as it is the one dealing with the problem of making sure someone receives the call. The function receives six different arguments. The first one is "$interval," which is the interval of time that it waits until repeating the call. The next four are the coordinates of an area on the screen that is monitored for its Pixel Sum, which by its turn is the sixth argument. Inside the function, the first instruction is the definition of the variable "$reached" as False. $reached is a Boolean, or logical, variable, which can be True or False. Thus, at the start of the function, we establish that the call did not reach its destiny. On the next line, we define the variable "$start" by making it receive the return of the function "TimerInit." This means that a timer is initiated at this moment, and the value of this timer is given later using the function "TimerDiff." The next line starts a While loop, which checks for $reached. It establishes that the loop will keep going while $reached is False, which so far is the case, since we set $reached to False at the start of the code. The next line contains an If … Then check, where the value of the timer initiated with $start is compared to the value of $interval. If it is smaller, then it calls another If … Then check, this time for the pixel sum of the passed coordinates. In this case, they correspond to the dial figure (Figure 6.4), which disappears when the phone is answered.

Therefore, if the dial symbol disappears, the sum of the pixels in the area changes, and this indicates that the phone was answered. Then, $reached receives True, which will terminate the While loop. However, if the call is not answered, the pixel sum will always be the same and the timer will keep increasing in value at each loop step (the pace is set at 100 ms, see the Sleep at the end of the loop). Thus, at a certain point, the timer will stop being smaller than $interval. When this happens, the Else of the first If … Then is reached, which means that the call is terminated using TerminateCallSkype, and a new call is started, using CallSkype, and the timer is reset, by redefining the $start variable. Then, the process repeats indefinitely, only really finishing if the phone call is answered.

The functions in Code sample 6.7 can be combined to automate the phone call and dealing with the error in FAKAS execution:

```
"Alarms.au3"
$Sent = False
While $Sent = False
    If WinExists("Neddle error", "") Then
        CallSkype(0266269565)
        RepeatUntilReachSkype(30000,400,400,445,440,758017266)
        SoundPlay("C:\WarningMessage.wma",1)
        TerminateCallSkype()
        $Sent = True
    EndIf
WEnd
```

Code sample 6.8 Automated call to a phone using Skype after an error during FAKAS execution.

Before running the script, make sure Skype is open. By running the script, you will see nothing happening unless the error window (Figure 6.1) in FAKAS exists. In this case, you should see Skype making a phone call to the selected number. If no one gets the phone call in 30 s, it terminates the call and starts again. If someone gets the phone call, then the function SoundPlay plays the recorded sound file (the argument "1" after the file path determines that the file is played entirely and only then the script proceeds). Analyzing the code, you see that a phone number will be called using CallSkype, and then the function that checks for the answer and repeats the call until a positive response (RepeatUntilReaachSkype) is called with the appropriate arguments. See that the time interval was set to 30 s. Once successful, the voice message is played. After that, the call is finished (TerminateCallSkype), $Sent receives True, and the loop (and consequently the script) is finished.

Because this script relies on pixel watching, you may consider using BlockInput (see Section 4.6) to ensure that the mouse and keyboards are locked while the script is running. However, since the script should in principle run while there is nobody near the computer, such measure is probably not necessary. You could also improve this script (Code sample 6.6) by adding a second phone number to be called in case the first one does not respond after 10 min or so, for example.

6.3 Summary

- AutoIt can easily automate the control of communication software and integrate it with analytical software, thus enabling the setup of simple alarms.
- It is very easy to create a function to work on a browser interface and send e-mails. Here, SendGmailOpera was created.
- It is also possible to send e-mails without using any browser by means of SMTP, which stands for Simple Mail Transfer Protocol.
- Using Skype, the instant messaging software, it is possible to send SMS.
- It is possible to send voice messages to phone lines that cannot receive SMS using software like Skype.
- The functions TimerInit and TimerDiff are used to set up timers.

7

Using Low-Cost Equipment for Laboratory Automation

The synchronization of programs shown in Chapters 3–5 does not need to be restricted to software that controls laboratory instruments. Nowadays, low-cost automated machines are becoming more and more common, and some of them can be useful in laboratories. In this chapter, we explore some of the options that are available and that can be suited for a wide range of tasks in a laboratory.

7.1 G-Code Devices

G-code is a programming language that has been used since the 1950s for industry automation. G-code is arguably simpler than ordinary computer programming and, for most purposes of laboratory automation, only the simplest aspects of the language are necessary.

Three-dimensional (3D) printers and computer numeric control (CNC) routers can be called "G-code devices" as they can be programmed using G-code. In addition to being potentially useful tools for laboratories on their own, CNC routers and 3D printers can easily be adapted to perform repetitive tasks common to many laboratory environments, like sampling, for example. Therefore, they can substitute purpose-made autosamplers that normally have a much higher cost. For these reasons, it is useful for technicians and scientists to at least be aware of these possibilities, so that this route can be taken at a very low cost and enable what could be an expensive (or very labor-intensive) project.

7.1.1 CNC Routers

CNC stands for computer numeric control, meaning that the machine (in this case, a router) is controlled by the computer. In most cases, the machine moves a drill on three perpendicular axes (Figure 7.1).

A CNC needs to be sturdy, because the precise drilling of hard materials is mechanically demanding. Nowadays, there are CNC routers under US$ 1000, and there are many resources on the Internet teaching how to build one from scratch, which means that you could have a working machine for even less money. However, for most people without good electronic and mechanical skills, buying an assembled unit can be a good idea, especially compared with the usual price of scientific equipment. The only thing that

---

Practical Laboratory Automation: Made Easy with AutoIt, First Edition. Matheus C. Carvalho.
© 2017 Wiley-VCH Verlag GmbH & Co. KGaA. Published 2017 by Wiley-VCH Verlag GmbH & Co. KGaA.

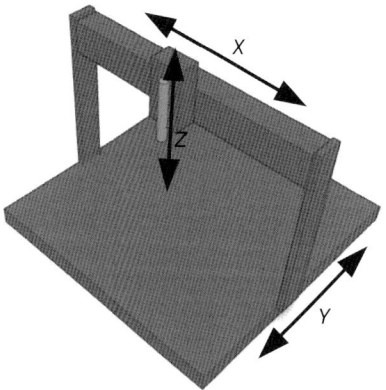

Figure 7.1 Hypothetical CNC router. In a typical CNC router, the drill is moved on the three perpendicular axes.

you would need to do would be to replace the drill with a more appropriate tool, like a syringe or needle, so that the CNC can be used to sample liquids, for example.

A very popular software to control CNC routers is Mach 3, by ArtSoft. It can be downloaded for free from http://www.machsupport.com/software/downloads-updates/. In this configuration, it is limited to 150 lines of code, which is very little for routine routing work but more than enough for adapting a CNC for laboratory automation. To remove this limitation, you can pay US$ 175 so that your code can have as many lines as you need. A problem with Mach 3 is that it is only compatible with older versions of Windows, like XP, and operates only through parallel (printer) ports. Therefore, it is becoming outdated. However, since there are still many Windows XP computers around, and Mach 3 is still widely used, it is worthy to cover it. In addition, the principles of Mach 3 operation are similar to those of other software, and thus can serve as an illustration. Install Mach 3. If you just follow the normal installation steps, several programs will be installed. Among them, you need to choose "Mach 3 Mill."

If you have bought a CNC, you need to ensure that Mach 3 can communicate with it, and you should consult the manual of the machine for specific information. Once your router is communicating with Mach 3, you should be able to control it by using the keyboard arrow keys. Before that, make sure to press the large RESET button so that the warning message is not displayed. Only after this, you can start moving your router. Using the keys, you can move the tool attached to the router on three axes: X, Y, and Z. Although it may be different in your case, let us assume here that the X-axis is the left–right direction, Y-axis the back–forth direction, and Z-axis the up–down direction (Figure 7.1). An important aspect you will probably note is that if you reach one of the extremes of the possible positions, for example, if you take the tool totally to the left, the motor will keep going but the tool will not move. Furthermore, the position on the software will keep increasing (or decreasing, depending on how you configured your axis). This is probably different from autosamplers that you may have dealt with.

After getting used to these characteristics of the CNC router, let us prepare it so that it can be useful for automation work. The first step is to define the origin (or zero) positions for each axis. You can choose any zero that you wish, but a common configuration is to move the tool totally to the back, to the left, then to the top, and after that setting this position as zero on the Mach 3 interface (See Figure 7.1, and find the buttons "Zero X," "Zero Y," and "Zero Z"; you should press these buttons). Now, if you press the arrow keys to move the tool, avoid reaching zero, because if you go beyond the limit, the coordinates will be meaningless.

7.1.2 G-Code for CNC

There are many resources teaching G-code, and only the simplest (but probably the most useful in the context of laboratory automation) commands will be presented here: G0 (fast linear movement) and G1 (slow linear movement). For example, G0 X30 Y40 Z10 means that the object of interest (e.g., the tip of a needle for water suction or delivery) will be moved at maximum speed to positions 30, 40, and 10 on a 3D coordinate system of three axes (X, Y, and Z). For G1, a typical line would be G1 X30 Y40 Z10 F100, which means the same movement previously described for the G0 example, but this time done at a controlled speed ("feed rate"; thus the code F) of 100 µm s^{-1}. Instructions are saved on a text file (extension .tap for Mach 3).

Let us work with an example. Suppose your CNC router is used as an autosampler, and there are three samples, all at the $X=0$ line and at $Y=0$, $Y=3$, and $Y=6$ (units are not important now, but on Mach3 you can set them to millimeter or inches). The sampling procedure consists in moving the needle from the initial $X=Y=Z=0$ position to one of the sample's position, bringing the needle down to $Z=-2$ (inside the vial), waiting for sampling to be finished, bringing up the needle to position $Z=0$, and finally bringing the needle back to position $X=Y=0$. These steps should be carried out in synchrony with the analytical software. The first step to enable this to work is to write the G-codes. Open Notepad, write and save in your script folder (the folder where you have been saving your scripts so far) four different .tap files with the following four codes:

```
TrayCNC-0.tap:
G1  Z0  F200
G0  X0  Y0
%

TrayCNC-1.tap:
G0  X0  Y0
G1  Z-2  F200
%

TrayCNC-2.tap:
G0  X0  Y3
G1  Z-2  F200
%

TrayCNC-3.tap:
G0  X0  Y6
G1  Z-2  F200
%
```

Each file has three lines. For file TrayCNC-0.tap, the first line determines that the tool moves vertically to position 0 at a speed of 200 µm s^{-1}. The second line of the file determines that the tool now goes to position $X=0$ and $Y=0$ at maximum speed. As you see, the vertical movement was made separately from the horizontal movement, which is important if you are controlling a needle that goes in a vial, for example. Finally, the third line with a % marks the end of the code. The other three files have a similar structure: first a quick horizontal movement, and then a slow vertical movement to position −2. Vertical movements were set to be slow because some needles are fragile when going through septa, and going slowly helps to avoid problems in some cases. However, you can certainly use G0 instead of G1 to do a fast movement in this step if you do not have such concern with your samples.

You can test the G-codes using Mach 3. Open the Mach 3 window, then choose the File menu and select "Open G-code." You should then select the folder where the files that you created were saved and then open them. Then, press Alt + r or click on the "Cycle Start" button. You should see the values near each axis in the "REF ALL HOME" block changing to the set values in the G-code. You should also see the movement tracker at the top right-hand side of the interface drawing the trajectory of the tool (needle). If you have a real CNC router, you will see that these G-codes represent very little movement in real life. If you change the values for X, Y, and Z in the files, and save them with another name, you can see larger movements.

7.1.3 Synchronizing a CNC Router to a Laboratory Instrument

After having the G-codes, we need to write the AutoIt scripts that will enable the integration of the CNC router with an analytical equipment. The first code that we need to write is the sequence of actions that the CNC router performs. This can be written as a function, and saved as a file named CNC-functions.au3:

```
Func CNCMoveToSample($posit)
   WinActivate("Mach3")
   Send("!f")
   Send("{ENTER}")
   Sleep(1000)
   ControlSetText("Open","","Edit1","TrayCNC-"&$posit&".tap")
   Sleep(1000)
   ControlClick("Open","","Button2")
   Sleep(1000)
   WinActivate("Mach3")
   Send("!r")
   Sleep(7*1000)
EndFunc
```

Code sample 7.1 Function CNCMoveToSample with the sequence of steps for sampling with the CNC router controlled using Mach3.

Let us see what the function CNCMoveToSample does. First, it activates the Mach3 window, and then sends Alt + f, which opens the File menu. Then it sends Enter, which opens a pop-up window that has "Open" as its title. It waits for 1 s, then types the name of the file to be open using the command ControlSetText. Then it waits for another second, and clicks the "Open" button using ControlClick. After that, it activates Mach3 window again, and sends Alt + r, which starts the software. Finally, it sleeps for 7 s.

As you can see, the actions in Code sample 7.1 are exactly those that you would do manually if controlling Mach3. Note that controls were not used at all steps, and that there were some sleeps between control commands. As explained in the chapter about controls (Chapter 5), not all software can be accessed using controls. This is the case of Mach3. If you open AWI (AutoIt v3 Windows Info), try to get control information for the "Cycle Start" button, for example. You will see that it is not available. Therefore, keyboard shortcuts were necessary. The sleeps between commands were set in order to ensure that commands would respond. Sometimes, the software delays its response to a command and if a delay is not set, the second command can be missed. Finally, the last sleep of 7 s allows for the movements of the CNC router to be finished. This function, then, mixes the techniques presented in several different chapters. It illustrates the real world of scripting: when possible, we use controls and interactive scripting; if not, we rely on arbitrary timing and keyboard shortcuts and, in the worst case, mouse clicks.

We can use the new function CNCMoveToSample together with those defined for Fake Carbon Analyzer Controller (FACACO) in the file FACACOFAKASfunctions.au3 and write a script synchronizing Mach3 and FACACO. The modified script is as follows:

```
#include "FACACOFAKASfunctions.au3"
#include "CNC-functions.au3"
opt("WinTitleMatchMode",1)
FACACOstartControl()
Sleep(2000)
For $sample = 1 to 3
   CNCMoveTosample($sample)
   FACACOmeasureControl($sample)
   CNCMoveTosample(0)
   FACACOstatusControl()
Next
```

Code sample 7.2 Script that integrates FACACO and Mach3.

Run the script with the usual precautions for FACACO (see Chapter 3). It should work in the same way the previous codes worked for FACACO and FAKe AutoSampler (FAKAS). Pay close attention and see that the position of the needle shown on the Mach3 interface is different reflecting the content of each G-code at each run, following the different G-codes called each time.

7.1.4 3D Printers

As you could see, the control of a CNC router as an autosampler using G-code and AutoIt is simple. Any other G-code device, as a 3D printer, can be used similarly. Differently from CNC routers, 3D printers do not drill through materials but, instead, create a shape by dropping very small drops of melted plastic. Therefore, 3D printers do not need to be sturdy, and can be quite light. They do need, however, a hot plate where the plastic is deposited, and an extruder that releases the plastic. Also, a typical 3D printer differs from most CNC routers in that the movements are divided between two parts of the printer: the extruder, which moves up and down, and the hot plate, which moves forward, backward, left, and right (Figure 7.2). An advantage of 3D printers over many CNC routers is that they are in almost all cases connected using a USB (universal serial bus) cable to the computer, which means that any modern computer can be connected to them.

Low-cost 3D printers are becoming very common, with many examples costing less than US$ 500. These cheaper machines, however, often come as kits that you need to assemble yourself. The cost can be even lower if buy parts separately from different

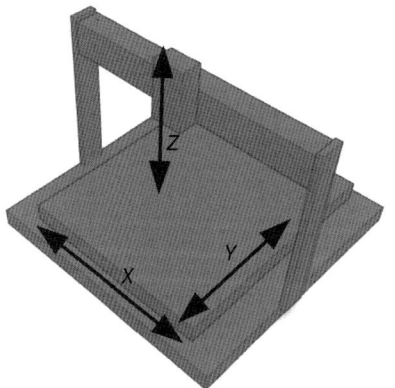

Figure 7.2 Hypothetical 3D printer. In a typical 3D printer, a horizontal plate moves on the horizontal axes and the extruder moves on the vertical one.

suppliers and build a printer from scratch (only recommended for people with good electronic and mechanical knowledge). As for CNC routers, it is necessary to adapt a needle or syringe replacing or at the extruder so that the printer can be used as an autosampler. There are several different programs that control 3D printers. As Mach3, they allow the input of G-code files, although the extension is not .tap in most cases. With this and other small differences in mind, you can easily adapt the examples shown for the CNC and write scripts to integrate such software to your analytical software.

7.2 Robotic Arms

An alternative to G-code devices are robotic arms. Nowadays, less expensive robotic arms are available (even less than US$ 100), which can be controlled by a computer and, as you already know, can be integrated with a laboratory instrument to perform a task, like working as an autosampler, for example. The main advantage of robotic arms over G-code devices is that robotic arms take much less space, and can be very light; it is possible, for example, to fix a robotic arm on a vertical wall and use it from there, if appropriate. Doing the same with a CNC router is very hard. The main disadvantage is that robotic arms are much harder to program than G-code devices, as we will see. Also, cheaper models tend to be fragile and thus cannot perform actions that demand strength above a certain level.

Some key points are important when choosing a robotic arm. First, the type of motor that moves them. Motors in the cheapest robotic arms are DC motors. Such motors have no position control. Therefore, the only way to program movements for such robotic arms is by means of timing how long a motor moves clockwise or in the opposite direction. Reproducibility of movements in such cases can be compromised if motors deteriorate, for example. Therefore, such type of robot is not recommended.

Most robotic arms use servo motors instead of DC motors for their movements. Servo motors allow precise position control, but their movements are restricted to a 180° arc. For robotic arms, this is not a problem. Servo motors can be controlled using open-source microcontrollers such as Arduino, and thus it is possible to build robotic arms like these from scratch. However, there are many kits for sale, several under US$ 500, which can be quite good to automate tasks in the laboratory. In fact, I have used one as an autosampler for several chemical measurements in water (see the suggested

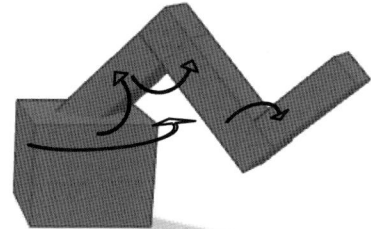

Figure 7.3 Hypothetical robotic arm. In most robotic arms, movements are semicircular, which make them more difficult to program than G-code devices.

bibliography at the end of the chapter). Despite successful, programming a robotic arm as described in my previous work is very time-consuming. The main reason is that the movements of a robotic arm are semicircular (Figure 7.3). It is more difficult to program movements based on angles or calibrating positions based on angles to then calculate points on the three perpendicular axes than to simply input these positions as it is possible with G-code devices.

Finally, "serious" robotic arms are available in the market at a price of above US$ 1000. These do not need to be built, and are fully tested in the factory. They, however, need to be programmed. The manufacturer provides the instructions for programming and, once you set up the movements, you can easily integrate it with analytical software using AutoIt, as shown for other cases in this book. If the robots are calibrated in factory, programming them can be almost as easy as for G-code devices. At the time of writing this book, new, exciting low-cost robotic arms were being built by start-ups on crowdfunding. The reader can profit greatly from keeping up to date on such developments, which, by means of scripting, can become a very useful and low-cost solution for laboratories.

7.3 Do-It-Yourself Devices

A recent development in manufacturing technology has been the adoption of open-source hardware, which consists of hardware that has its blue prints available for all. With the availability of microcontrollers and manufacturing equipment like 3D printers, building open-source hardware is more accessible than ever. This has led to the development of numerous tools that substitute those normally available for scientists at a much lower cost, and potentially with the same or even superior performance, since custom-built instruments become more accessible.

The number of open-source laboratory instruments is increasing rapidly, and the reader is encouraged to explore the bibliography about the topic listed at the end of the chapter. If the intention is to integrate such open-source devices using AutoIt, however, it is important to remember that they ideally should be controlled using Windows, and not other operating systems (see Section 2.6).

7.4 Summary

- With AutoIt, it is possible to incorporate low-cost devices as parts of an automation setup in a laboratory.
- G-code devices (3D printers and CNC routers) are very simple to control, and can be of low cost readily adapted to perform tasks in a laboratory.

- CNC routers are sturdy and usually have a fixed base.
- Three-dimensional (3D) printers are generally less sturdy than CNC routers and often have a movable base.
- Robotic arms are more difficult to program than G-code devices, because robotic arms move their parts in semicircles, not straight lines.
- Open-source hardware has become increasingly available, and can also be integrated with other laboratory equipment using AutoIt.

Suggested Reading

G-code:

You can find many sources on the Internet, which for the basic purposes outlined here are more than sufficient. Simply search for G-code plus 3D printer or CNC using your favorite search engine.

Open-source hardware in laboratory automation:

http://www.journals.elsevier.com/hardwarex

Carvalho, M.C. and Eyre, B.D. (2013) A low cost, easy to build, portable, and universal autosampler for liquids. *Methods Oceanogr.*, **8**, 23–32.

Gross, B.C., Erkal, J.L., Lockwood, S.Y., Chen, C., and Spence, D.M. (2014) Evaluation of 3D printing and its potential impact on biotechnology and the chemical sciences. *Anal. Chem.*, **86**, 3240–3253.

McMorran, D., Chung, D.C.K., Li, J., Muradoglu, M., Liew, O.W., and Ng, T.W. (2016) Adapting a low-cost selective compliant articulated robotic arm for spillage avoidance. *J. Lab. Autom.*, in press.

Pearce, J.M. (2012) Building research equipment with free, open-source hardware. *Science*, **337**, 1303–1304.

Pearce, J.M. (2014a) Cut costs with open-source hardware. *Nature*, **505**, 618.

Pearce, J.M. (2014b) *Open-Source Lab: How to Build Your Own Hardware and Reduce Research Costs*, Elsevier.

Symes, M.D., Kitson, P.J., Yan, J., Richmond, C.J., Cooper, G.J.T., Bowman, R.W., Vilbrandt, T., and Cronin, L. (2012) Integrated 3D-printed reactionware for chemical synthesis and analysis. *Nat. Chem.*, **4**, 349–354.

8

Arrays and Strings

Before proceeding with the "applied" contents of this book, it may be necessary to make some readers familiar with two ways of organizing data that are used in programming: strings and arrays. Readers who have a background in programming can skip this chapter, or may give a quick look, only to learn the name of the functions that deal with them. Otherwise, it is suggested to spend some time practicing the scripts provided here, until you fully grasp the concepts.

So far in this book, we have dealt with data that consist of a single value. Variables received values as numbers or words, or parts of words (characters). However, sometimes it is useful to deal with sets of values. In this chapter, we will see how AutoIt deals with organized data (arrays) and raw data (strings).

8.1 Organized Data: Arrays

Arrays are sets of data. The main utility of using arrays is that we can organize data that are somehow related to each other in a group. In addition, many AutoIt functions are based on arrays, and thus it is essential to understand how they work. Let us create an array containing only numbers:

```
Dim $Array1[5] = [1,2,3,4,5]
MsgBox(0,"Elements",$Array1[1])
```

Code sample 8.1 Creating an array containing five numbers and displaying the second one in the list.

Running the script, you see that "2" is displayed in the message box. You can see in Code sample 8.1 that 2 is the second element in the array, but that it was called as being "$Array1[1]." This is because the first element in an AutoIt array is indicated by [0]. You must keep this in mind. Also, your array must be declared using either "Dim," "Local," or "Global" as a modifier, unless they are the receiving end of a function returning data. This is because when creating an array we must allocate the memory that will be used. Finally, note that by changing the number inside the [] you can access any element in the array, as long as the number is lower than the array length. This is a fundamental property that makes arrays especially useful to recover the necessary data from a set.

Let us see another example, now an array of words:

```
Dim $Array1[5] = ["red","green","pink","blue","white"]
For $i = 0 to 4
   MsgBox(0,"Elements",$Array1[$i])
Next
```

Code sample 8.2 Creating an array containing five words and displaying all of them.

Running the script, you will see the five words that are elements of the array being displayed. Note that because the first element is [0], the loop starts at 0 and goes up to 4, instead of 5. If you use 5 as the upper limit for the loop, you will get an error when reading the last element ($Array[5]), because it will not exist.

Arrays like those in Code samples 8.1 and 8.2 are the simplest type of array, the unidimensional arrays, that is, data are listed on a single line. It is possible to create arrays of up to 64 dimensions in AutoIt. However, rarely we need more than two dimensions. Bi-dimensional arrays can be seen as very similar to spreadsheets, for example. See the following code for an example:

```
Dim $Array1[2][2] = [[1,2],[4,5]]
For $i = 0 to 1
   For $j = 0 to 1
      MsgBox(0,"Elements",$Array1[$i][$j])
   Next
Next
```

Code sample 8.3 Creating and displaying a bi-dimensional array.

The array created in Code sample 8.3 has four data, organized into two columns and two rows.

In addition to creating arrays, AutoIt can do many operations with them, such as sorting, transposing, or multiplying. However, for the examples provided in this book, we will not use these operations. Interested readers can study the Array.au3 user-defined function (UDF), which includes many functions dealing with arrays.

8.2 Raw Data: Strings

Strings are sequences of characters. Examples of strings are a single word; single number; sequence of words; mixture of words and numbers; any mix of words, numbers, empty spaces, and other characters like #$; and so on. Strings in AutoIt can have up to 2 147 483 647 characters, which is like a book with more than 750 000 pages. Therefore, it is unlikely that you will ever need to worry about having a string that is too long.

It is important to deal with strings, because they can be interpreted as a series of data. When you exchange information between computers, or between computers and machines, you exchange data, which almost never come ready and polished for direct use. One way to deal with such data is to treat them as a string, and use the functions that enable their processing. The procedure of dealing with raw data is called *parsing*, and is an important field in computer science. In this book, only the most basic techniques of parsing are presented. AutoIt has several functions dealing with strings. Let us see some simple ones:

```
$String1 = "The quick brown fox jumps over the lazy dog"
$String_length = Stringlen($String1)
MsgBox(0,"String length",$String_length)
$First5totheleft = StringLeft($String1,5)
MsgBox(0,"First 5 characters:",$First5totheleft)
$UpperCase = StringUpper($String1)
MsgBox(0,"All in upper case:",$UpperCase)
```

Code sample 8.4 Some basic string functions.

Remember that strings always come between quotes (""), as in line 1 in this code. Running the script, you should see message boxes displaying the number of characters of "$String," which is 43, the "First 5 characters" of the string (The q) and the whole string in upper case (THE QUICK BROWN FOX JUMPS OVER THE LAZY DOG). As you see, you can get information about the string, get a piece of it, and create a new one based on it. There are other functions that complement the ones presented here, you can find them in the AutoIt help file.

For purposes of exchanging data between computers, a very useful function is "StringSPlit," which converts a string into an array and thus enables the several parts of the string to be accessed individually. We will use StringSplit in several of the next scripts in this book. Let us see an example:

```
$String1 = "The quick brown fox jumps over the lazy dog"
$StringArray = StringSplit($String1," ")
MsgBox(0,"Fourth element:",$StringArray[4])
```

Code sample 8.5 Introducing StringSplit.

Run the script, and you should see a window displaying the fourth word of the phrase that composes $String1, which is "fox." The arguments for StringSplit are the string to be split and " ", that is, an empty space. It means that StringSplit will split the string at the empty spaces, creating an array that contains each word, as defined by the characters contained between empty spaces, as an element in this case. You can change the separating character from empty space and use "o," for example, and you will get a very different result in your message box.

If you paid attention to the Array section, you may find it strange that "$StringArray[4]" returned the fourth (fox), and not the fifth (jumps), word in the string. As you remember, arrays in AutoIt start at [0]. What happens here is that StringSplit creates an array, which has the first element containing the number of words in the string, and the first word goes to [1]. It is confusing, and must be remembered when using this function.

Another function that can be useful when using StringSplit and working with the resulting array is "Ubound." Ubound returns the number of elements of an array:

```
$String1 = "The quick brown fox jumps over the lazy dog"
$StringArray = StringSplit($String1," ")
MsgBox(0,"0th element:",$StringArray[0])
MsgBox(0,"Array length:",Ubound($StringArray))
```

Code sample 8.6 Introducing Ubound.

Running the script, you will see two message boxes, the first displaying 9 and the second 10. Therefore, the first one displays the number of "real" elements in the array, that is, excluding the [0], and Ubound displays the total number, including [0].

As a final word about strings, many advanced and complex procedures (parsing) can be done with them, which can be useful if the incoming data set to be analyzed is complex. Regular expressions are the approach used in such cases. AutoIt can deal with Regular Expressions, but this subject will not be covered here. Interested readers can find plenty of information on regular expressions on the Internet, and even books devoted solely to this subject.

8.3 Summary

- Arrays and strings are "collective" data types, that is, they consist of more than one element.
- Elements in arrays are numbered. In AutoIt, the first element has index [0].
- Arrays can have more than one dimension.
- If you create a new array, you need to declare it at some point in the code. If the array is created to receive the output of a function, you do not need to declare it before that.
- Elements in strings are not numbered. Therefore, it is more difficult to deal with them than with arrays.
- Strings are often the type of data that instruments transmit to computers. Therefore, it is very important to learn how to process them.
- Strings can be converted to arrays using the function StringSplit.
- The array created using StringSplit differs from ordinary arrays in AutoIt, because its first element contains the number of subsequent elements in the array.

9

Data Processing with Spreadsheets

Computer-aided laboratory activities are not limited to operating machines and getting measurements done. After measurements are finished, there is the need for processing the data and distributing the results for laboratory clients. In this chapter, we will see how AutoIt can be used to make some of these postanalysis tasks easier.

9.1 Exporting Results to Spreadsheet Software

9.1.1 Selecting Spreadsheet Software

It is very common in laboratories the use of spreadsheet software to process data and distribute results. In this chapter, we will see how AutoIt can be used to automate the export of measurement results to spreadsheet software. Two different spreadsheet programs will be covered: Microsoft Excel and LibreOffice Calc. These two programs are chosen because they have already libraries of functions written for them in AutoIt. Furthermore, they are widely popular and Excel can effectively be considered the standard spreadsheet software being used by the largest portion of users. Calc, on the contrary, is less widespread, but entirely free, which Excel is not, and is compatible with Excel anyway, which means that you can use Calc and your files can be read by someone using Excel. Some readers may be missing OpenOffice Calc. Unfortunately, the library available for AutoIt only works for LibreOffice, and thus OpenOffice will not be covered in this chapter. However, one of the techniques presented will work for this program too, and should work for virtually any modern spreadsheet program.

Some readers may know that both Excel and Calc allow the automation of their procedures by means of their built-in scripting languages. However, it is much easier to use AutoIt than such languages to integrate the software with other programs, which makes AutoIt a better choice in the context of laboratory automation, because the software controlling instruments is not Excel or Calc. Another advantage is that many scripts written here are version-independent, that is, they work for any version of Excel or Calc. This is not the case for the scripting languages that come as standard for such software: sometimes, with a new software version the script language is also modified, and thus scripts that were written for older versions may not work in newer ones.

You might be wondering if automation of Excel and Calc is done by means of controls. In fact, it is not. It is a different type of automation, which is not covered in this book. It is based on the "application program interface" (API) of the software being

Practical Laboratory Automation: Made Easy with AutoIt, First Edition. Matheus C. Carvalho.
© 2017 Wiley-VCH Verlag GmbH & Co. KGaA. Published 2017 by Wiley-VCH Verlag GmbH & Co. KGaA.

automated. Some software manufacturers publish the API for their programs, so that other programs can interact with them. This is one of the approaches used by instrument manufacturers to allow compatibility between their software and spreadsheet software like Excel. It demands much more technical skills than the ones covered in this book, and is one example that illustrates how complicated it was to integrate products made by different manufacturers before AutoIt became available.

Now that the technical background has been covered, let us get to some more practical matters. In order that the scripts presented in this book work, it is necessary that you have Excel or Calc installed on your computer. Excel 2013 and Calc version 5.0.2 were tested. Remember: Calc is entirely free and can be downloaded, together with the remaining of the LibreOffice package, from https://www.libreoffice.org/download/libreoffice-fresh/. Therefore, you should be able to test these scripts without any problem.

9.1.2 Transferring Data to Spreadsheets

Perhaps the most common role of spreadsheet software in laboratory automation is the processing of results generated by other programs. Therefore, a tool to create an Excel file from the results obtained from the software is very useful, and in fact present in many products as a standard feature. However, in some cases, such feature is not present and, instead, a text (.txt) or comma separated value (.csv) file is created. In some cases, such files are created with the purpose of being opened later using Excel or Calc, and are structured in an organized way, which could be, for example, separating results from other data using always the same character, like a space or comma, and using line separators. Other times, however, the program does not generate any export file. It is impossible to generalize, and hence it will be shown how this can be done with Fake Carbon Analyzer Controller (FACACO). By studying the list of examples provided in this chapter, you may get inspired to create an ideal solution for your particular case.

In our first example, let us see how we can get data generated on the spot and transfer it to Excel or Calc bypassing the generation of .TXT or .CSV files. Let us see how this could be done for FACACO:

```
1) #include <Excel.au3>
2) opt("WinTitleMatchMode",1)
3) $ProgExcel = _Excel_Open()
4) $workbook = _Excel_BookNew($ProgExcel)
5) For $sample = 1 to 9
6)    $result = ControlGetText("FAk","","Edit"&30+$sample)
7)    _Excel_RangeWrite($workbook, $workbook.Activesheet, $result, "A"&$sample)
8) Next
```

Code sample 9.1 Getting results (first nine lines) from FACACO and sending to Excel.

Remove the line numbers and, before running the script, fill the Result fields of FACACO with any numbers of your choice (example in Figure 9.1). Make sure you have Excel installed on your computer and then run the script. You should see the results from FACACO being exactly copied to the first column of the spreadsheet on Excel.

Let us now see the new components in Code sample 9.1. The library "Excel.au3" is initially included. This library contains "_Excel_Open," "_Excel_BookNew," and

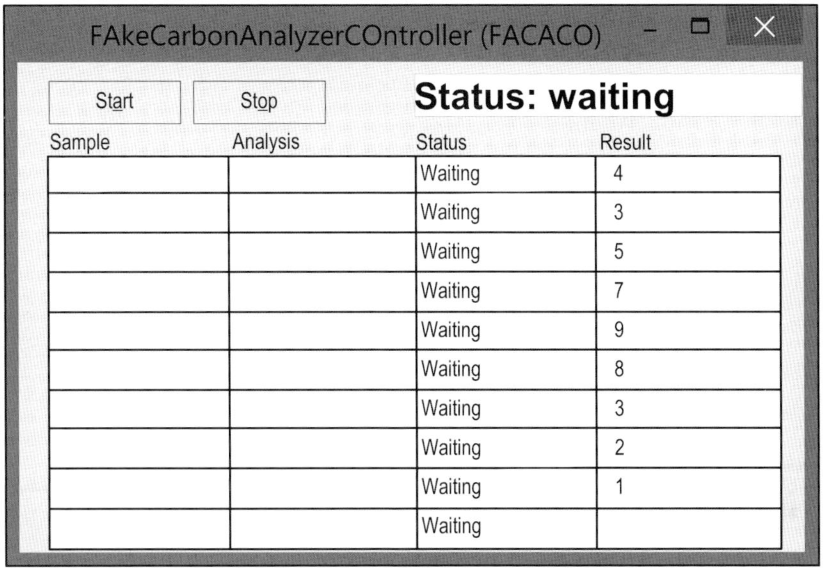

9.1 Exporting Results to Spreadsheet Software

FAkeCarbonAnalyzerCOntroller (FACACO)

Sample	Analysis	Status	Result
		Waiting	4
		Waiting	3
		Waiting	5
		Waiting	7
		Waiting	9
		Waiting	8
		Waiting	3
		Waiting	2
		Waiting	1
		Waiting	

Status: waiting

Figure 9.1 FACACO filled with hypothetical results.

"_Excel_RangeWrite," among many other functions. These three functions follow very closely what their names indicate: _Excel_Open starts Excel, assigning it to the "$ProgExcel" variable. _Excel_BookNew creates a new Excel file using $ProgExcel as an argument, and assigning the "$workbook" variable with it. Finally, _Excel_RangeWrite writes the value of the variable "$result," which was the contents of a result field in FACACO (which changes for every loop step) to the open excel file, $workbook, and, among its spreadsheets, the active spreadsheet, which is indicated by the argument "$workbook.Activesheet." The value of $result is written from cell A1 to A9, as indicated by the last argument of the function.

The same procedure can be done for LibreOffice Calc. First, you need to download the library from https://www.autoitscript.com/forum/topic/118280-debugged-enhanced-udf-files-for-open-office/ and copy the file "OooCalc.au3" to the folder in which you are running your script.

```
#include <OOoCalc.au3>
opt("WinTitleMatchMode",1)
$ProgCalc = _OOoCalcBookNew()
For $sample = 1 to 9
    $result = ControlGetText("FAk","","Edit"&30+$sample)
    _OOoCalcWriteCell($ProgCalc, $result, $sample-1,0)
Next
```

Code sample 9.2 Getting results (first nine lines) from FACACO and sending to Calc.

As for Code sample 9.1, make sure you have some values in the Result fields in FACACO. By running the script, you should see them being transferred to a Calc spreadsheet. As in Code sample 9.1, the main new features are the inclusion of the necessary library (in this case, "OooCalc.au3"), and two new functions: "_OooCalcBookNew," and "_OooCalcWriteCell." _OooCalcBookNew both opens Calc and creates a new spreadsheet, assigning it to "$ProgCalc." _OooCalcWriteCell writes to $ProgCalc the desired value ("$result") at the desired cell, which is defined by the coordinates "$sample-1" (row) and 0 (column), meaning that the results will be pasted at different rows from 0 to 8 (1-1 to 9-1), and on column 0. In Calc, the first line and column are labeled 0 and not 1. Thus, a big difference in relation to the Excel version is that here the notation "A1" is not used, but instead the numbers relative to each cell position. A disadvantage of the Calc library compared to that for Excel is that SciTE does not support autocomplete for Calc.

What if you do not use either Excel or LibreOffice Calc and prefer another spreadsheet program? In this case, it is still possible to export data automatically. This solution is general and works for Excel and LibreOffice Calc as well.

```
opt("WinTitleMatchMode",1)
WinActivate("Untitled")
Send("^{HOME}")
For $sample = 1 to 9
   $result = ControlGetText("FAk","","Edit"&30+$sample)
   WinActivate("Untitled")
   Send($result)
   Send("{DOWN}")
Next
```

Code sample 9.3 Getting results (first nine lines) from FACACO and sending to a generic spreadsheet program.

Take the same precautions as those for the previous examples in this chapter and run the script. You should see the data being transferred from FACACO to your spreadsheet (provided its window title starts with "Untitled"; if not, change the script to suit your case). The code is quite simple without any new elements if you have been following the book sequence so far.

An even more general solution is possible for programs for which controls are not accessible. Suppose this were the case for FACACO. In this case, a working code could be:

```
opt("WinTitleMatchMode",1)
WinMove("FAke","",0,0)
WinActivate("FAke")
MouseClick("left",300,270)
WinActivate("Untitled")
Send("^{HOME}")
For $sample = 1 to 9
   WinActivate("FAke")
   Send("{TAB}")
   Send("^c")
   WinActivate("Untitled")
   Send("^v")
   Send("{DOWN}")
Next
```

Code sample 9.4 Getting results (first nine lines) from FACACO without using controls and sending those to a generic spreadsheet program.

Again, take the precautions that you took for the previous scripts in this chapter, and run the script. As in the other cases, you should see the data being transferred from FACACO to the spreadsheet. This time, however, no controls were accessed in FACACO, and only copy (Ctrl + c) and paste (Ctrl + v) were used. Again, no new elements were present in this script. It may be worth commenting on the strategy, though. After moving FACACO to a fixed (0,0) position, a mouse click is applied to the lowest cell in the Status column. This was done so that by pressing TAB, the next cell to be chosen would be the first cell in the Result column, and automatically its contents would be selected. This made it much easier to write the rest of the script dealing with FACACO. When writing a similar script, it is worth paying attention to such properties of the programs being automated so that scripting becomes easier.

9.1.3 Transferring Data in Real Time

There are cases in which measurement results are not available to be saved. For example, ahead in this book, we write a program that controls a balance and keeps showing the weight constantly (Code sample 20.6). It is possible to deal with these situations and transfer the data to spreadsheets, if so desired. Suppose that FACACO would not leave the results of the measurements for each sample being displayed "forever" on its window. In such a case, you would need to copy the results as soon as they are generated. Let us see the code to do that for FACACO:

```
1)  #include <Excel.au3>
2)  opt("WinTitleMatchMode",1)
3)  $ProgExcel = _Excel_Open()
4)  $workbook = _Excel_BookNew($ProgExcel)
5)  For $sample = 1 to 3
6)      While ControlGetText("FAk","","Edit"&30+$sample) = ""
7)          Sleep(20)
8)      WEnd
9)      $result = ControlGetText("FAk","","Edit"&30+$sample)
10)     _Excel_RangeWrite($workbook, $workbook.Activesheet, $result, "A"&$sample)
11) Next
```

Code sample 9.5 Getting FACACO results as they are generated and sending them to Excel.

This time, take the usual precautions that you took in the initial chapters (see details in Section 3.5), and remove the line numbers. Run the script and you will see that every time FACACO finishes a sample, the result of the measurement is transferred to Excel. This was enabled by adding the While loop that verifies the contents of the result fields in FACACO (they must be all clear at the start) and, when not empty, transfer them to Excel. The same strategy could easily be used for Calc, by adding the same loop to Code sample 9.2, or for a generic program, adopting the relevant changes in Code samples 9.4 and 9.5. For even faster data retrieval, for example, in cases in which a sensor coupled to the computer keeps generating data constantly, see Section 17.3.

9.2 Dealing with Saved Results (Files)

Some analytical programs store data in proprietary formats. Sometimes it is necessary to deal with such files, in some occasions with many of them. For example, it is common that after a month, or a year, there is the need to do some accounting of which types of measurements were done, how many, and for whom. In many laboratories, such tasks are well organized. However, there is always the chance that an unexpected request comes and then some data mining is required. If you are familiar with such a scenario, know that AutoIt can make the task easier.

When searching for the results of previous chemical analyzes, it can be necessary to examine the files of a given format that were created during a given time. There are some good tools, including free tools, that do the job for Windows. However, AutoIt can provide more flexibility, and hence we can use the functions together with other programs, including spreadsheet and database software. This can lead to very convenient and powerful solutions for file management.

Perhaps the most basic task in file management is to list the files in a folder. AutoIt has a library (File.au3) with the function "_FileListToArray," which does this job. See an example:

```
#include <file.au3>
#include <OOocalc.au3>
$path = "C:\Program Files (x86)\AutoIt3"
$FileListShort = _FileListToArray($path,"*",1)
$FileListLong = _FileListToArray($path,"*",1,True)
$Book1 = _OOoCalcBookNew()
For $i = 1 to $FileListShort[0]
   _OOoCalcWriteCell($Book1,$FileListShort[$i],$i-1,0)
   _OOoCalcWriteCell($Book1,$FileListLong[$i],$i-1,1)
Next
```

Code sample 9.6 Script introducing _FileListToArray.

Note that you will need to have LibreOffice Calc installed on your machine and that the library Ooocalc.au3 must be in the same folder of your script so that it will work properly. You can replace these lines using Excel functions or using the other approach for a generic spreadsheet, both techniques shown in the same chapter. Taking these precautions, run the script. You should see two lists of files being written to the Calc spreadsheet. The first, to the left, contains file names. The one next to it contains the full file name, including the path. Both data sets were created using the same function, _FileListToArray, which was used twice to create two arrays: "$FileListShort" and "$FileListLong" (check Chapter 8 for an explanation on arrays). When creating $FileListShort, the arguments for _FileListToArray were "$path," which was the path to the folder of interest, "*," which means all the files (it is assumed that you are familiar with common Windows wildcards), and 1. If you check the help file for this function, you will see that selecting 1 in the third argument you establish that only files, and not folders, will be listed. If you want to include folders, this could have been left blank, as it is the default. For $FileListLong, the difference was that after 1 the argument "True" was used. Also in the help file, you will see that setting True here you get the full path of the files, as opposed to only their names, in the standard configuration ("False").

Another function to list files is "_FileListToArrayRec." This function is a more powerful version of _FileListToArray, the main difference being that it can list the files in subfolders inside the folder of interest. It is possible to apply very sophisticated filters in order to get exactly the files that you want. See the following example:

```
#include <file.au3>
#include <OOocalc.au3>
$path = "C:\Program Files (x86)\AutoIt3"
$FileListRecShort = _FileListToArrayRec($path,"*|*.au3;*.exe",1,1,0,0)
$FileListRecLong = _FileListToArrayRec($path,"*|*.au3;*.exe",1,1,0,2)
$Book1 = _OOoCalcBookNew()
For $i = 1 to $FileListRecShort[0]
    _OOoCalcWriteCell($Book1,$FileListRecShort[$i],$i-1,0)
    _OOoCalcWriteCell($Book1,$FileListRecLong[$i],$i-1,1)
Next
```

Code sample 9.7 Script introducing _FileListToArrayRec.

As for the previous example, you will need to have Calc installed on your computer, or will need to make the changes as explained for the previous code. Run the script, and you will see that now other files are listed, different from those obtained with Code sample 9.6. As for this previous code, two columns of the spreadsheet are filled with the file lists. The first list shows only file names and the second one shows the full path. As for _FileListToArray, this difference comes from the argument passed for _FileListToArrayRec. In this case, it is the last argument: 0 for only file names and 2 for the full path. The other arguments are: $path, which is the path to the desired folder; "*|*.au3;*.exe," which says that we want all files (the first *) except (what comes after |) those of au3 and exe type; 1, which determines that only files, and not folders, will be listed (check the help file), another 1, which enables recursion, that is, it says that subfolders will be included in the list; and finally a 0, determining that files will not be sorted in any way (the following argument, 0 or 2, was already explained). Note that _FileListToArray and _FileListToArrayRec are functions that organize the input, originally strings, to arrays, like StringSplit, introduced in Chapter 8.

Not impressed with these functions? True; so far nothing very useful that many free utilities cannot do better. So let us do something now that would be difficult to do without AutoIt. First, you will need to copy the following files from http://www.wiley-vch.de/publish/en/books/ISBN978-3-527-34158-0/. They are six .jpg files, three of them with red shapes and the other three with blue shapes. The shapes are square, circle, and triangle (Figure 9.2).

Figure 9.2 Files with different figures available from http://www.wiley-vch.de/publish/en/books/ISBN978-3-527-34158-0/ and that should be saved in the C:\PrettyPics folder. It is difficult to see as the figure is in gray scale, but first, fourth, and fifth are in blue, and the others in red in the original files.

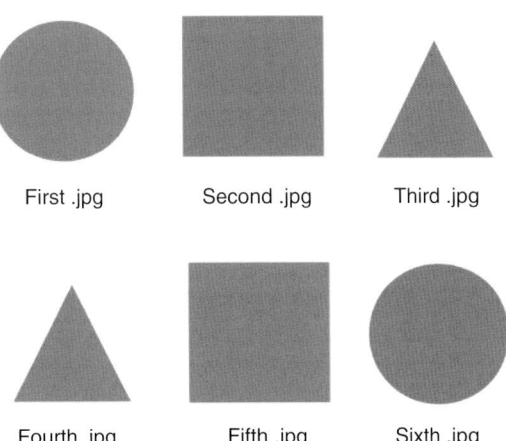

Save these files to a new folder (PrettyPics, in the C:\folder). Also, you will need a image processing software. I personally like IrfanView, which is free and available from http://www.irfanview.com/main_download_engl.htm. It will be used in this example. The script below lists only the files containing blue figures:

```
#include <file.au3>
#include <OOocalc.au3>
opt("WinTitleMatchMode",2)
$path = "C:\PrettyPics"
$FileListShort = _FileListToArray($path,"*.jpg",1,True)
$Book1 = _OOoCalcBookNew()
$line = 0
For $i = 1 to $FileListShort[0]
    ShellExecute($FileListShort[$i])
    Sleep(1000)
    WinMove("jpg","",0,0)
    If PixelGetColor(315,160) = 6710782 Then
        _OOoCalcWriteCell($Book1,$FileListShort[$i],$line,0)
        $line = $line+1
    EndIf
    Sleep(1000)
    WinClose("jpg")
    Sleep(1000)
Next
```

Code sample 9.8 Organizing figure files according to color.

In order that the script works, you will need to make sure that IrfanView is the default software set to open .jpg files. If not, you will need to modify the script to suit the program that you are using. Running the script, you will see that IrfanView will be called (by the function "ShellExecute") and closed (function WinClose) six times, each for each jpg file in the folder. When doing that, AutoIt checks the pixel color at position 315,160 using the function PixelGetColor (more details in Section 4.2). If the color is blue (color code 6710782, see the procedure to obtain such color code in Section 4.2), the file name is typed on the first column of the spreadsheet. See that, differently from the previous two scripts, now the line where the data are typed (indicated by the variable "$line") increments only if there are data to be typed. This avoids the column of the spreadsheet becoming full of empty lines.

This example was very simple, but illustrates the potential of the technique for automating the selection of files that would be impossible using traditional means. In a laboratory, the same technique could, for example, be used to select only files that had valid runs, which would be clear by the absence or presence of a given graphical pattern or numerical value in the result file. Such an example has not been shown, because it can be very specific, and its principles will be the same of Code sample 9.8.

Now let us see a semiautomated procedure to list the files according to the shape of the figure in them. Different from pixel color, it is not trivial to get the shape of a figure in a picture file. The same situation may arise with instrument-generated data, which might need user's evaluation that cannot be replaced by an automated procedure. You may be thinking that this negates the purpose of using AutoIt. However, this is not true,

because the procedure as a whole is much faster than manually checking, than copying
and pasting file names and so on. See the code below:

```
#include <file.au3>
#include <OOocalc.au3>
opt("WinTitleMatchMode",2)
$path = "C:\PrettyPics"
$FileListShort = _FileListToArray($path,"*.jpg",1,True)
$Book1 = _OOoCalcBookNew()
$line0 = 0
$line1 = 0
$line2 = 0
For $i = 1 to $FileListShort[0]
    ShellExecute($FileListShort[$i])
    $shape = InputBox("Shape","s: square, t: triangle and c: circle")
    Switch $shape
        Case "s"
            _OOoCalcWriteCell($Book1,$FileListShort[$i],$line0,0)
            $line0 = $line0+1
        Case "t"
            _OOoCalcWriteCell($Book1,$FileListShort[$i],$line1,1)
            $line1 = $line1+1
        Case "c"
            _OOoCalcWriteCell($Book1,$FileListShort[$i],$line2,2)
            $line2 = $line2+1
    EndSwitch
    Sleep(1000)
    WinClose("jpg")
    Sleep(1000)
Next
```

Code sample 9.9 Organizing figure files according to shape.

Run the script. Again, you will need to have IrfanView installed, and the figures in the correct folder. This time, when IrfanView is called, also an input box comes and you need to classify the image. You are asked to type c, t, or s and, depending on your choice, the file is grouped in one of the three different columns selected for the output in the Calc spreadsheet.

In addition to making lists of files, AutoIt enables other useful file operations that can make it easier some tedious tasks that are sometimes part of the work in a laboratory. The interested reader should consult the help file and explore the functions listed in File.au3.

9.3 Processing Spreadsheet Files

In some occasions, it can be necessary to process spreadsheet files containing the data from several previous runs. This is not difficult to do manually, but it can be tedious and time-consuming. Let us see how using spreadsheet and file functions provided by AutoIt can make such a task easier. First, let us create 40 Calc files (you will need to have LibreOffice Calc on your computer; if you prefer Excel, you will need to modify the scripts using Excel functions from excel.au3). These files will contain the results of some hypothetical measurements. The code below will create the files:

```
1)  #include <OOoCalc.au3>
2)  opt("WinTitleMatchMode",1)
3)  For $books = 1 to 40
4)      $Book = _OOoCalcBookNew()
5)      $line = 0
6)      $linestd=0
7)      $linesample=0
8)      $lineblank=0
9)      _OOoCalcWriteCell($Book, "Sample type", $line,0)
10)     _OOoCalcWriteCell($Book, "Sample name", $line,1)
11)     _OOoCalcWriteCell($Book, "Result", $line,2)
12)     _OOoCalcWriteCell($Book, "OBS", $line,3)
13)     For $line = 1 to 30
14)         If $line = 1 or $line = 7 or $line = 15 or $line = 22 or $line = 30 Then
15)             $linestd = $linestd + 1
16)             _OOoCalcWriteCell($Book, "Standard", $line,0)
17)             _OOoCalcWriteCell($Book, "Book"&$books&"Standard"&$linestd, $line,1)
18)             _OOoCalcWriteCell($Book, Random(24,25), $line,2)
19)         ElseIf $line = 2 or $line = 29 Then
20)             $lineblank = $lineblank + 1
21)             _OOoCalcWriteCell($Book, "Blank", $line,0)
22)             _OOoCalcWriteCell($Book, "Book"&$books&"Blank"&$lineblank, $line,1)
23)             _OOoCalcWriteCell($Book, Random(0,0.5), $line,2)
24)         Else
25)             $linesample = $linesample + 1
26)             _OOoCalcWriteCell($Book, "Sample", $line,0)
27)             _OOoCalcWriteCell($Book, "Book"&$books&"Sample"&$linesample, $line,1)
28)             _OOoCalcWriteCell($Book, Random(3,50), $line,2)
29)             $samplevalue = _OOoCalcReadCell($Book, $line,2)
30)             If $samplevalue > 45 Then
31)                 _OOoCalcWriteCell($Book, "Over limit", $line,3)
32)             EndIf
33)         EndIf
34)     Next
35)     WinActivate("Untitled")
36)     Send("!f")
37)     Send("a")
38)     Sleep(1000)
39)     Send("C:\PrettyPics\Book"&$books)
40)     Sleep(1000)
41)     Send("{ENTER}")
42)     Sleep(1000)
43)     Send("^q")
44)     Sleep(1000)
45) Next
```

Code sample 9.10 Script that creates 40 Calc spreadsheets in a specific folder.

Remove the line numbers, run the script, and now you will see Calc being called 40 times, each time 31 lines being written to each spreadsheet, and then the file being saved and closed. The only new functions in this code are "_OOoCalcReadCell," which returns the value of a cell to a variable (in this case, "$samplevalue") and "Random," which generates pseudo-random numbers in a predefined range (the arguments are the lower limit and upper limit of this range). By now, if you have been following the book, you should be able to understand the code without difficulty. Still, in order to make the task easier, a short description will be given, as the code is relatively long. It consists of a big loop that calls actions to be repeated 40 times, each time creating a new .ods file. The variable "$line" refers to the lines of the spreadsheet. Column titles are given on line 0 ($line = 0, see _OooWriteCell being called at the start of the code). The next three variables ("linestd," "linesample," and "lineblank") are used to fill the cells with relevant data, that is, for example, standard1 or blank2. The sub-loop from 1 to 30 is where the cells are filled with data. In all cases, some lines are reserved for standards and blanks, and the rest goes for samples. The Random function is called to provide the values for each measurement. For blanks, the value is set to be between 0 and 0.5, for the standard between 24 and 25, and for the samples any value between 3 and 50. Finally, if the value given to a sample is larger than 45, an observation is made that it was over the limit of correct measurement. After the data are typed, several Send commands are sent in order to save and close the file.

By running Code sample 9.10, you have 40 files in your folder. The next step is to extract only the relevant data. Suppose these relevant data are only samples, not standards or blanks. Suppose also that invalid results, that is, those above the measurement limit, are not interesting either. Suppose you want a list of all these relevant data inside a single spreadsheet file. The following script does this job:

```
1)  #include <file.au3>
2)  #include <OOocalc.au3>
3)  opt("WinTitleMatchMode",2)
4)  $path = "C:\PrettyPics"
5)  $FileListShort = _FileListToArray($path,"*.ods",1,True)
6)  $BookDestiny = _OOoCalcBookNew()
7)  $LineDestiny=0
8)  _OOoCalcWriteCell($BookDestiny, "Sample type", $LineDestiny,0)
9)  _OOoCalcWriteCell($BookDestiny, "Sample name", $LineDestiny,1)
10) _OOoCalcWriteCell($BookDestiny, "Result", $LineDestiny,2)
11) For $i = 1 to $FileListShort[0]
12)     $BookSource = _OOoCalcBookOpen($FileListShort[$i])
13)     For $LineSource = 2 to 30
14)         $sampletype = _OOoCalcReadCell($BookSource, $LineSource,0)
15)         $samplename = _OOoCalcReadCell($BookSource, $LineSource,1)
16)         $sampleresult = _OOoCalcReadCell($BookSource, $LineSource,2)
17)         $sampleobs = _OOoCalcReadCell($BookSource, $LineSource,3)
18)         If $sampletype = "Sample" Then
19)             If $sampleobs <> "Over limit" Then
20)                 $LineDestiny = $LineDestiny + 1
21)                 _OOoCalcWriteCell($BookDestiny, $sampletype, $LineDestiny,0)
22)                 _OOoCalcWriteCell($BookDestiny, $samplename, $LineDestiny,1)
23)                 _OOoCalcWriteCell($BookDestiny, $sampleresult, $LineDestiny,2)
24)             EndIf
25)         EndIf
26)     Next
27)     _OOoCalcBookClose($BookSource)
28) Next
```

Code sample 9.11 Script transferring selected data from 40 spreadsheet files to a single destination file.

Run the script, you will see that a new spreadsheet is created, then each one of those 40 files that were created using the previous script are open and their data are selected according to the criteria outlined in the previous paragraph. The new functions in this code are "_OOoCalcBookOpen" and "_OooCalcBookClose," which, as their name imply, open and close .ods files using Calc. This script uses _FileListToArray to gain access to all .ods files in the specified folder. Then, a new spreadsheet is created using _OooCalcBookNew, and some column titles are typed on line 0. After that, a loop starts opening one of the previously created .ods files for each run. The open file has its contents repeatedly transferred (by another loop) to variables using _OooCalcReadCell, and after that these contents are either tested for some values or transferred to the destiny spreadsheet. Each file is closed after their contents have been used, and the loop continues until the last file.

Other more complex types of data operations are possible using approaches similar to this. Moreover, database software could be integrated into the script and even more complex data mining could be possible. The next chapter introduces a way of working with databases in AutoIt.

9.4 Summary

- AutoIt has libraries containing functions that make it very easy to input and read data from major spreadsheet software packages, such as Excel and Calc.
- Alternatively, it is simple to use other software by using other AutoIt functions.
- An AutoIt library, File.au3, contains several useful functions that deal with file management.
- _FileListToArray lists all files in a folder by making an array. There are many options for this function.
- _FileListToArrayRec is a more powerful function that enables more sophisticated file listing.
- By combining file and spreadsheet functions, very efficient data handling becomes possible.

10

Working with Databases

In the Chapter 9, you learned how to automate some aspects of data export and processing using spreadsheet software. This type of approach will work for many situations, but there are cases in which database software work better. For example, when your dataset is very large, or when you plan to do complex data grouping and analysis, database software is easier to use than spreadsheets. Also, most modern database software packages deal with relational databases, which allow very sophisticated data organization and analysis. It is true that most computer users never tried database software, or, if they tried, they returned to their spreadsheets. A common reason is that spreadsheets are straightforward to use, while databases are not, requiring some training for their proper use. However, due to the characteristics of database software, some users could benefit from learning how to use them. In this chapter, I am not going to teach about databases; there books entirely for that purpose. Instead, I simply present a simple introduction to SQlite, which has a user-defined function (UDF) defined for AutoIt, and is a very popular database library. Although at first sight, it may look more difficult than using Access (Microsoft Office database program) or Base (LibreOffice database program), SQlite is not more difficult than them for the uninitiated user.

10.1 Starting SQlite in AutoIt

AutoIt comes with a UDF to utilize SQlite features. It can be called by including the library "SQlite.au3." However, this is not enough to get starting. You also need to download the proper .dll file, otherwise your scripts will not work. You can download it from http://https://www.sqlite.org/download.html, where you should look for "Precompiled Binaries for Windows." The version that you should get is the 32-bit DLL (×86). You need to download the corresponding file, uncompress it, and then, copy the files that composed the compressed files to your system folder, which you can discover by simply typing MsgBox(0,"",@SystemDir) and running as a script on SciTE. It is possible that new versions come after this book is published. Therefore, it is a good idea to test whether the correct files were downloaded. You can do this using this script, which is a simplified version of a very similar one in the AutoIt help file:

Practical Laboratory Automation: Made Easy with AutoIt, First Edition. Matheus C. Carvalho.
© 2017 Wiley-VCH Verlag GmbH & Co. KGaA. Published 2017 by Wiley-VCH Verlag GmbH & Co. KGaA.

```
#include <SQLite.au3>
$sSQliteDll = _SQLite_Startup()
If @error Then
   MsgBox(0, "SQLite Error", "SQLite3.dll not found!")
   Exit -1
EndIf
MsgBox(0, "SQLite3.dll Loaded", $sSQliteDll)
_SQLite_Shutdown()
```

Code sample 10.1 Checking whether the correct DLL file is being accessed. If yes, its path will be shown. If not, an error message will be displayed.

If you copied the correct files to you system folder, its path will be shown. If not, an error message will be displayed. If you get an error, try another file from those listed among the "Precompiled Binaries for Windows" on the SQlite website.

Code sample 10.1 introduces some essential functions to work with SQlite in AutoIt. The first one is "_SQLite_Startup." This function loads the SQlite3.dll file, the one that you downloaded. Without this, nothing can be done in SQlite, and thus this function must be called at every script dealing with SQlite. The other is "_Sqlite_Shutdown," which unloads the .dll file. Although you can run your scripts without including this function, because AutoIt automatically closes any .dll at the end of the scripts, it is recommended that you include it in your code.

10.2   Creating SQlite Databases

Once you have the .dll file to start SQlite, you gain access to the full SQlite language and can create, modify, and consult databases. Let us first see how we create a simple database:

```
#include <SQLite.au3>
_SQLite_Startup ()
$DBname = InputBox("File name","Input file name(anything.db)")
$DB = _SQLite_Open($DBname)
_SQLite_Exec($DB, "CREATE TABLE TABLE1 (Car, Weight int)")
_SQLite_Exec($DB, "BEGIN")
_SQLite_Exec($DB, "INSERT INTO TABLE1 VALUES ('honda',850)")
_SQLite_Exec($DB, "INSERT INTO TABLE1 VALUES ('mercedes',1250)")
_SQLite_Exec($DB, "INSERT INTO TABLE1 VALUES ('toyota',950)")
_SQLite_Exec($DB, "COMMIT")
_SQLite_Close()
_SQLite_Shutdown()
```

Code sample 10.2 Creating a simple SQlite database.

If you run Code sample 10.2, a window will appear asking you to type a file name. You can choose any file, and here the example will be with a file named "cars.db." After that, the script finishes, and that is all. If you go to the folder where you saved your script, you will find cars.db. Before analyzing the code in detail, it can be good to be able to open the database that was created, so that explanations become clearer. In order to do that, you need a program that can read SQlite files. A very good one is SQlite browser, which can be freely downloaded from http://sqlitebrowser.org. Install the program, and start it. On SQlite browser window, send Ctrl + O, and navigate until you find the file "cars.db." You need to change the type of file being searched for to "All files" to be able to find it. Once you open the file, information about it will be displayed.

Once we can look at our database, let us examine Code sample 10.2. Three new functions are introduced in this code: "_SQlite_Open," "_SQlite_Exec," and "_SQlite_Close." _SQlite_Open, as its name implies, opens an existing database. However, it can also be used to create a new one. Here, the database was stored in a file, whose name was provided by the user in the InputBox (see how the variable that received this name, "$DBname," was used as an argument for _SQlite_Open). The variable "$DB" received the return of _SQlite_Open, and was used as an argument for _SQlite_Exec, which is the function that creates tables and input data, among other features.

_SQlite_Exec received as arguments the database to be worked with and an SQlite instruction, which is based on SQL (structured query language). Note, for instance, that the instructions passed as arguments are not highlighted in SciTE, as they are not part of AutoIt language, but of SQL. Not all SQlite instructions that are possible to be passed for the different functions will be covered in this chapter. There is a large number of instructions, and you should consult an SQlite specialized text if you want to become familiar with them. In this chapter, only a few of the most common ones will be presented. In its first instance, _SQlite_Exec received as instruction "CREATE TABLE TABLE1 (Car, Weight int)"; CREATE TABLE is the instruction that creates a new table, which here was named TABLE1. As you may know, a database is composed of one or more tables. For example, if you check Figure 10.1, you will see that there is a space for tables in the "Database Structure" tab, and that TABLE1 is listed there.

TABLE1 had two columns: Car and Weight. Car is treated as text, because it was not modified, but Weight is a number, more specifically an integer, as it was modified by "int" (meaning integer). On SQlite browser, choose the "Browser Data" tab (Figure 10.1) and there you will see the columns, together with the data that were input using _SQlite_Exec in the following lines of the code (Figure 10.2).

_SQlite_Exec was called five other times (Code sample 10.2). In one of them, the argument was BEGIN, basically stating that from that point on the database would be modified. After that, _SQlite_Exec was called thrice, in which variations of "INSERT INTO TABLE1 VALUES" were used as arguments. INSERT INTO is the instruction that allows the insertion of data in a table, and VALUES are the data to be inserted. See that for each line of code, different values were entered, but always names for cars

| Database Structure | Browse Data | Edit Pragmas | Execute SQL |

Figure 10.1 Options available for database management using SQlite browser.

Figure 10.2 SQlite browser displaying TABLE1, a part of cars.db database.

	Car	Weight
	Filter	Filter
1	honda	850
2	mercedes	1250
3	toyota	950

and integer numbers for weights. If you want, you can, for example, change one of the numeric values in the arguments for a word, and run the script. The console (see Section 2.5.2) will display an error message, and the database will not be properly created, with the wrongly input data missing. Note also that text inputs came within quotes (e.g., " " and ' '), and the numeric ones did not.

Finally, _SQlite_Exec was called once more, this time with the COMMIT argument, stating that operations in the database would be finished. After that, the other new function, _SQlite_Close, was called to close the database file.

The example given in Code sample 10.2 was obviously merely illustrative. It is pointless to use AutoIt to create a database in that way, because it is much easier and more intuitive to do the same job directly on SQlite Browser (you may want to explore the program and learn how to do it, it is straightforward). Let us see now an example in which creating the database using AutoIt is easier than using SQlite Browser:

```
1) #include <SQLite.au3>
2) $path = "C:\PrettyPics"
3) $FileListShort = _FileListToArray($path,"*.ods",1,True)
4) _SQLite_Startup ()
5) $DBname = InputBox("File name","Input file name(anything.db)")
6) $DB = _SQLite_Open($DBname)
7) _SQLite_Exec($DB, "CREATE TABLE TABLE2 (File_name,
File_size_in_bytes int, Date_created_YYYY_MM_DD)")
8) _SQLite_Exec($DB, "BEGIN")
9) For $i = 1 to $FileListShort[0]
10)    $name = $FileListShort[$i]
11)    $size = FileGetSize($name)
12)    $ArrayCreated = FileGetTime($name,$FT_CREATED)
13)    $created =
$ArrayCreated[0]&"-"&$ArrayCreated[1]&"-"&$ArrayCreated[2]
   _SQLite_Exec($DB, "INSERT INTO TABLE2 VALUES
('"&$name&"','"&$size&"','"&$created&"');")
14) Next
15) _SQLite_Exec($DB, "COMMIT")
16) _SQLite_Close()
17) _SQLite_Shutdown()
```

Code sample 10.3 Creating an SQlite database for the several files in a folder, displaying their path, size, and date of creation.

In order to properly run this script, you must first run the one in Code sample 9.10, because that script creates the files that are used as input here. Once you have done so, run the script, and in a few instants your database will be created, displaying the path, size, and date of creation for each .ods file in the folder. The file-related functions were presented in Chapter 9, so please have a look there to better understand them. Let us focus on the SQlite part here. Comparing to Code sample 9.2, there are no new SQlite functions. What is different here is that the arguments passed as SQlite instructions to _SQlite_Exec now contain variables. Note that variables must be input within ' " & & " '. Be very careful with this notation, otherwise the script will not work properly.

At this point, you should have no problems creating SQlite databases using AutoIt. If you wish, you could try to adapt the scripts in Chapter 9, including the one that organized files using a semiautomated method for files containing different geometrical shapes.

10.3 Modifying an Existing SQlite Database

It is a very common situation that, once your database contains a table, you realize that you need to add new columns to it. While this can very easily be done manually using SQlite browser (go to Database structure, Figure 10.1, click on an existing table, and then on "Modify Table"), with AutoIt we can automate the task.

Suppose, for example, that we want to associate our list of spreadsheets to the clients for the measurements. As you remember from Section 9.3, each table created there corresponds to the results of a series of measurements. The idea is to link each spreadsheet to its respective client. In this case, we will suppose there were eight clients, each corresponding to five spreadsheet files. To make it simple, since the intention here is only to present the technique, the files are already organized in the correct order.

An obvious way would be to simply rewrite Code sample 10.3 from scratch and simply add a new column and the appropriate data there. For a very small database as the one with what we are dealing here, this is a feasible solution. However, for realistic databases containing tens of thousands of entries, this is not an efficient solution. Let us see how we can simply add a new column to a database without needing to change anything in the existing data:

```
#include <SQLite.au3>
_SQLite_Startup ()
$DBname = InputBox("File name","Input database to be modified")
$DB = _SQLite_Open($DBname)
_SQLite_Exec($DB, "ALTER TABLE TABLE2 ADD COLUMN Client_name")
_SQLite_Close ()
_SQLite_Shutdown ()
```

Code sample 10.4 Adding a new column to the database created in Code sample 10.3.

Run the script, and open the file using SQlite browser. Now, you should find an extra column among those in TABLE2. However, you will also see that there are no entries in this column; we will deal with this soon. Now, just look at the code. The only new element is the instruction passed to _SQlite_ Exec as an argument: "ALTER TABLE TABLE2 ADD COLUMN Client_name." You might be noting a pattern here, when you compare with the previous code samples in this chapter. If so, you may have guessed that ALTER TABLE was the command modifying TABLE2, and that ADD COLUMN was the specific modification, in this case, the addition of the column "Client_name." The next step is to add the desired data to our new column, which consist of the names of the clients for each spreadsheet containing measurement data:

```
1)  #include <SQLite.au3>
2)  _SQLite_Startup ()
3)  $DBname = InputBox("File name","Input file name (anything.db)")
4)  $DB = _SQLite_Open($DBname)
5)  _SQLite_Exec($DB, "BEGIN")
6)  For $i = 1 to 40
7)     If $i < 6 Then
8)        $client = "James"
9)     ElseIf $i < 11 Then
10)       $client = "Mary"
```

```
11)        ElseIf $i < 16 Then
12)            $client = "Antonio"
13)        ElseIf $i < 21 Then
14)            $client = "Makiko"
15)        ElseIf $i < 26 Then
16)            $client = "Shu"
17)        ElseIf $i < 31 Then
18)            $client = "Muhammad"
19)        ElseIf $i < 36 Then
20)            $client = "Mpemba"
21)        Else
22)            $client = "Nadir"
23)        EndIf
24)     _SQLite_Exec($DB, "UPDATE TABLE2 SET Client_name =
'"&$client&"' WHERE ROWid = '"&$i&"'")
25) Next
26) _SQLite_Exec($DB, "COMMIT")
27) _SQLite_Close()
28)  SQLite_Shutdown()
```

Code sample 10.5 Adding data to the new column to the database created in Code sample 10.4.

Remove the line numbers, run the script, choose the file that contains your database, and, once it is finished, open the file using SQlite browser. Now you should see the new column, Client_name, all populated with data. The SQL instruction that populated the column with data was "UPDATE TABLE2 SET Client_name = '"&$client&"'" WHERE ROWid = '"&$i&"'," called as an argument for the _SQLite_Exec function. UPDATE acted on TABLE2, the table containing the data. SET acted on the column that we wanted to modify. Finally, WHERE allowed us to choose which lines of the column were to be modified. These lines are identified by "ROWid." See that the loop went from 1 to 40, because we knew that there were 40 lines.

You can use the UPDATE ... instruction presented here to modify any of the cells in the database. For example, suppose it was decided that the client "Nadir" was no longer responsible for the samples that were run, but, instead, now there is a new client named "Carlos." The next script shows how the data can be updated:

```
1) #include <SQLite.au3>
2) _SQLite_Startup ()
3) $DBname = InputBox("File name","Input file name(anything.db)")
4) $DB = _SQLite_Open($DBname)
5) _SQLite_Exec($DB, "BEGIN")
6) $client = "Carlos"
7) _SQLite_Exec($DB, "UPDATE TABLE2 SET Client_name = '"&$client&"'
WHERE Client_name = 'Nadir'")
8) _SQLite_Exec($DB, "COMMIT")
9) _SQLite_Close()
10) _SQLite_Shutdown()
```

Code sample 10.6 Changing the name of a single client in the database.

Run the script, and check the file using SQlite browser, as usual. You should see all instances of "Nadir" replaced by "Carlos." Note that there is no explicit loop in Code sample 10.6. This is because the repetitive action is done by SQLite, and not AutoIt.

Another common modification in databases is the removal of data. In the following example, all lines containing "Carlos" as Client_name will be removed. Before running the script, save a copy of the file where your database is stored with a different name, as we may need the original one for further examples.

```
#include <SQLite.au3>
_SQLite_Startup ()
$DBname = InputBox("File name","Input file name(anything.db)")
$DB = _SQLite_Open($DBname)
_SQLite_Exec($DB, "BEGIN")
$client = "Carlos"
_SQLite_Exec($DB, "DELETE FROM TABLE2 WHERE Client_name = 'Carlos'")
_SQLite_Exec($DB, "COMMIT")
_SQLite_Close()
_SQLite_Shutdown()
```

Code sample 10.7 Removing all lines containing "Carlos" as Client_name.

The SQL instruction that removes data is "DELETE FROM TABLE2 WHERE Client_name = 'Carlos'." The structure should be familiar by now: DELETE FROM acts on TABLE2, and WHERE on the column Client_name, choosing the lines that contain Carlos as their contents for deletion.

10.4   Databases with More Than One Table

Databases can have more than one table. In fact, this is one of the most important features of databases created by SQlite, that is, they are able to deal with many tables in a logical and useful way. In other words, SQlite deal with relational databases, which is a technical term for which the definition given in the previous sentence is only partially accurate. Relational databases can be structured in very complex ways, and this subject will not be explored, as it is vast, and could fill an entire book. The purpose here is to show how to create a database with more than one table, which is essential for a relational database. That being said, it is recommended to study the relational databases if you plan to use databases to organize your data.

Databases can be created from scratch with several tables, or new tables can be added to an existing database. Let us, for example, add a table to our database as it was after Code sample 10.6, that is, before removing the data as in Code sample 10.7.

```
#include <SQLite.au3>
_SQLite_Startup ()
$DBname = InputBox("File name","Input file name(anything.db)")
$DB = _SQLite_Open($DBname)
_SQLite_Exec($DB, "CREATE TABLE TABLE1 (Client_name,  Fund_source)")
_SQLite_Exec($DB, "BEGIN")
_SQLite_Exec($DB, "INSERT INTO TABLE1 VALUES ('James','Agency1')")
_SQLite_Exec($DB, "INSERT INTO TABLE1 VALUES ('Mary','Agency2')")
_SQLite_Exec($DB, "INSERT INTO TABLE1 VALUES ('Antonio','Agency3')")
_SQLite_Exec($DB, "INSERT INTO TABLE1 VALUES ('Makiko','Agency3')")
_SQLite_Exec($DB, "INSERT INTO TABLE1 VALUES ('Shu','Agency1')")
_SQLite_Exec($DB, "INSERT INTO TABLE1 VALUES ('Muhammad','Agency1')")
_SQLite_Exec($DB, "INSERT INTO TABLE1 VALUES ('Mpemba','Agency2')")
_SQLite_Exec($DB, "INSERT INTO TABLE1 VALUES ('Carlos','Agency1')")
_SQLite_Exec($DB, "INSERT INTO TABLE1 VALUES ('Nadir','Agency3')")
_SQLite_Exec($DB, "COMMIT")
_SQLite_Close()
_SQLite_Shutdown()
```

Code sample 10.8 Adding a new table to the database containing the previous data.

Run the script and choose the file that we have been using so far for our database as the file to have the new table, TABLE1, included. Opening it in SQlite browser, you will find both tables there. Note that one of the columns of TABLE1 is the same as the one in TABLE2, and it contains common elements. You could be thinking that it would be better to add a new column to TABLE2. However, as you can see, there are data in TABLE1 that are not in TABLE2: the client named Nadir. As you can imagine, it can be useful to keep Nadir as a client in a list, but not necessarily have her in a list of specific clients for a given measurement. Other users could be added to TABLE1, and could never be added to TABLE2, as another example.

10.5 Retrieving Data from Databases

The feature that makes databases powerful for data mining and organization is queries. SQlite is based on SQL, which is an acronym for structured query language. As you can imagine, queries are an extremely important part of SQL (they are the "Q" in SQL). Very sophisticated queries can be created using SQlite. In this section, only some very simple ones will be presented. The reader should consult specialized texts on the subject to grasp its full potential.

Let us create our first query. For that, we will need to have a database file created using Code sample 10.3, that is, containing the paths, sizes, and dates of creation for all .ods files in the C:\prettypics folder. Our first query will retrieve only paths of the files that are larger than 25 500 bytes:

```
1)  #include <SQLite.au3>
2)  Local $MyQuery, $DataArray
3)  _SQLite_Startup ()
4)  $DBname = InputBox("File name","Input file name")
5)  $DB = _SQLite_Open($DBname)
6)  _SQLite_Query($DB, "SELECT File_name FROM TABLE2 WHERE
    File_size_in_bytes > 25500", $MyQuery)
7)  While _SQLite_FetchData($MyQuery,$DataArray) = $SQLITE_OK
8)      ConsoleWrite($DataArray[0]&@CRLF)
9)  WEnd
10) _SQLite_QueryFinalize($MyQuery)
11) _SQLite_Close()
12) _SQLite_Shutdown()
```

Code sample 10.9 Query of the database of measurement files for files larger than 25 500 bytes.

Make sure you ran Code sample 10.3 and created a file, whose name you know before you run this script. You will need to input that file as the database to be queried. Here, the new functions are "_SQlite_Query," "_SQlite_FetchData," and "_SQlite_QueryFinalize." _SQlite_Query receives three arguments: the database to be queried, the SQlite instruction, and the variable that receives the result of the query, in this case, "$MyQuery." The instruction was "SELECT File_name FROM TABLE2 WHERE File_size_in_bytes > 25000." The first part of the instruction is SELECT, which selects the column of the table defined after FROM. Then, WHERE defines the condition that the value in the column File_size_in_bytes must be larger than 25 500.

10.5 Retrieving Data from Databases

A While loop is started, in which the result of the function _SQlite_FetchData is compared to the constant $SQLITE_OK. _SQlite_FetchData takes the query previously generated, $MyQuery, and creates an array based on it, "$DataArray." If the operation is successful, _SQlite_FetchData returns $SQLITE_OK. When there are no more data to be processed, the operation by _SQlite_FetchData returns a value different from $SQLITE_OK, and then the loop stops. Inside the loop, the values in the created array are printed in the console.

As the first example for database creation in the previous section (Code sample 10.2), Code sample 10.9 is more easily done using SQlite browser: just open the file with the database and select the "Execute SQL" tab. There, you can type your query (SELECT File_name FROM TABLE2 WHERE File_size_in_bytes > 25000), as in Figure 10.3. It should return the same results of the script in Code sample 10.9.

Now let us see an example that shows how using AutoIt and SQlite can make a task easier. Suppose that we need to make copies of the files of the original folder containing the measurement data (C:\Prettypics) to folders related to the clients of the measurements. See the code below:

```
1) #include <SQLite.au3>
2) Local $MyQuery, $DataArray
3) _SQLite_Startup ()
4) $DBname = InputBox("File name","Input file name")
5) $DB = _SQLite_Open($DBname)
6) Dim $ClientArray[8] =
["James","Mary","Antonio","Makiko","Shu","Muhammad","Mpemba","Carlos"
]
7) For $i = 0 to 7
8)     _SQLite_Query($DB, "SELECT File_name FROM TABLE2 WHERE
Client_name = '"&$ClientArray[$i]&"'", $MyQuery)
9)     DirCreate("C:\Prettypics\"&$ClientArray[$i])
10)    While _SQLite_FetchData($MyQuery,$DataArray) = $SQLITE_OK
11)       FileCopy($DataArray[0],"C:\Prettypics\"&$ClientArray[$i])
12)    WEnd
13)    _SQLite_QueryFinalize($MyQuery)
14) Next
15) _SQLite_Close()
16) _SQLite_Shutdown()
```

Code sample 10.10 Copying the files of the original measurement folder to folders divided according to client of the measurements.

Run the script and then open the C:\Prettypics folder. You should find eight new folders, each of them named after a different client. If you open the folders, you will see that

SELECT File_name **FROM TABLE2 WHERE** File_size_in_bytes **>** 25500

    File_name

1 C:\PrettyPics\Book1.ods

7 Rows returned from: SELECT File_name FROM TABLE2 WHERE File_size_in_bytes > 25500 (took 2ms)

Figure 10.3 Executing a query on SQlite browser that gives the same results as those from the script in Code sample 10.9. Only the initial part of the results is shown in the figure.

each of them contains five files, which are those files associated with each client in the database. The new functions in this code are "DirCreate," which creates a new folder, and "FileCopy," which copies the files from a folder to another. This example demonstrated how we can use AutoIt to combine SQlite and file management. Finally, let us see another example in the same line, but a little more complex:

```
1)  #include <SQLite.au3>
2)  Local $Query1, $Query1Array, $Query2, $Query2Array
3)  _SQLite_Startup ()
4)  $DBname = InputBox("File name","Input file name")
5)  $DB = _SQLite_Open($DBname)
6)  Dim $FundArray[3] = ["Agency1","Agency2","Agency3"]
7)  For $i = 0 to 2
8)      DirCreate("C:\Prettypics\"&$FundingArray[$i])
9)      _SQLite_Query($DB, "SELECT Client_name FROM TABLE1 WHERE Fund_source = '"&$FundArray[$i]&"'", $Query1)
10)     While _SQLite_FetchData($Query1,$Query1Array) = $SQLITE_OK
            _SQLite_Query($DB, "SELECT File_name FROM TABLE2 WHERE Client_name = '"&$Query1Array[0]&"'", $Query2)
11)         While _SQLite_FetchData($Query2,$Query2Array) = $SQLITE_OK
12)             FileCopy($Query2Array[0],"C:\Prettypics\"&$FundArray[$i])
13)         WEnd
14)         _SQLite_QueryFinalize($Query2)
15)     WEnd
16)     _SQLite_QueryFinalize($Query1)
17) Next
18) _SQLite_Close()
19) _SQLite_Shutdown()
```

Code sample 10.11 Copying the files of the original measurement folder to folders divided according to funding agency.

Run the script, and now you should see that new folders were created in the C:\Prettypics folder, but now for the different funding agencies. Opening the folders, you find the corresponding files for the measurements funded by each agency. There are no new elements in the code; it was a simple extension of Code Sample 10.10, in which a query was made inside another query.

10.6 Summary

- Databases can be a better option than spreadsheets when the purpose is to organize and access data.
- AutoIt has a library that supports SQlite, which is a popular database platform.
- The first step to work with SQlite is to download the right .dll file. You can check if you have downloaded the correct file by writing a simple script.
- The functions _SQlite_Startup, _SQlite_Open, _SQlite_Close, and _SQlite_Shutdown need to be called every time you work with SQlite on Autoit.
- _SQlite_Exec is one of the most important functions when working with SQlite on AutoIt. It is used to create and add new elements to a database.
- Although _SQlite_Exec is a single function, its arguments can be many, and include SQL.

10.6 Summary

- Throughout the chapter, several SQL instructions are presented. However, learning SQL is necessary if you aim to work seriously with SQlite.
- It is very helpful to work with a program that opens SQlite databases. A very good one is SQlite browser.
- SQlite deals with relational databases. Relational databases have more than one table and these tables share common data. Relational databases are very efficient to manage data.
- Data are retrieved from databases by means of queries.
- The functions that deal with queries are _Sqlite_Query, _SQlite_FetchData, and _SQlite_QueryFinalize.
- As for _SQlite_Exec, _SQlite_Query can receive many different SQL instructions as arguments.
- It is possible to open more than a query at once in a single script.

11

Simple Remote Synchronization

In Chapters 3–5, we learned how to synchronize instruments controlled by a single computer. However, it is not always possible to have all instruments being used to perform synchronized tasks connected to the same computer. Also, in some situations, it could be useful to have synchronized equipment working at different locations. In this and Chapters 12–16, some techniques that allow the synchronization between instruments controlled by different computers are presented. As you will see, all these techniques (except the one in Chapter 14) are quite simple and do not involve any knowledge of network protocols.

11.1 Time Macros

A very simple way to synchronize computers without any connection between them is by using time macros. Time and date are accessed in AutoIt by means of macros. Macros are special variables that cannot have a value assigned to them by the user; instead, they get their value from the Windows operating system. They are marked with a @, instead of $. See the example below:

```
$TargetHour = InputBox("Target hour","Please type the hour:")
$TargetMinute = InputBox("Target minute","Please type the minute:")
While $TargetMinute <> @MIN Or $TargetHour <> @HOUR
   sleep(1000)
WEnd
MsgBox(0,"","Done")
```
Code sample 11.1 Scheduling actions using the macros @HOUR and @MIN.

Run the script. You should see two input boxes asking for the hour and then the minute. Then, a While loop starts determining that while the minute or the hour input are different from the current hour (@HOUR) and minute (@MIN) the script will wait for 1 s and check them again. After that, the message "Done" appears to indicate that the script finished. Note that @HOUR is in the 24-h format, that is, 9 p.m. has a value of 21, and not 9.

Practical Laboratory Automation: Made Easy with AutoIt, First Edition. Matheus C. Carvalho.
© 2017 Wiley-VCH Verlag GmbH & Co. KGaA. Published 2017 by Wiley-VCH Verlag GmbH & Co. KGaA.

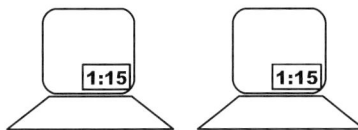

Figure 11.1 Scripts based on time macros are more easily built if the different computers being synchronized have exactly the same time being displayed.

There are other time macros in AutoIt, like @MDAY (day of the month), @YEAR (year), and @SEC (second), among others. A full reference is available in the help file of AutoIt.

11.2 Synchronizing FACACO and FAKAS Using Time Macros

Suppose that FACACO is being controlled by a computer, and FAKAS by another. In order to synchronize these instruments using time macros, the only prerequisite is that both computers have their clocks showing exactly the same time, preferably to the same second (but for many applications a 1 to 2 s difference will not be a problem; Figure 11.1). Here, in our examples, the scripts will work on the same computer, and thus the clock time will be the same. However, the reader can test on two separate computers and, if they have the same clock time setup, the scripts should work properly.

Since it is assumed that the instruments are connected to two different computers, it is necessary that two scripts are written, one for each computer. As in the previous chapters, we are going to write a function in order to make the code more organized:

```
Func WaitRightTime($hour, $minute)
    While $minute <> @MIN Or $hour <> @HOUR
        sleep(1000)
    WEnd
EndFunc
```

Code sample 11.2 Function that keeps the computer waiting until the proper clock time.

Copy this function to FACACOFAKASfunctions.au3 and save the file. Now, write the following scripts:

```
#include "FACACOFAKASfunctions.au3"
opt("WinTitleMatchMode",1)
WinMove("FAke","",0,0)
$hour = InputBox("Hour for FACACO","Please input hour:")
$minute = InputBox("Minute for FACACO","Please input minute:")
WaitRightTime($hour,$minute)
FACACOstartShortcut()
For $sample = 1 to 3
    Sleep(11500)
    FACACOmeasure()
    Sleep(5*1000)
Next
```

Code sample 11.3 Script to control FACACO on a computer assuming that the script in Code sample 11.4 will be used on another computer controlling FAKAS.

```
#include "FACACOFAKASfunctions.au3"
opt("WinTitleMatchMode",1)
WinMove("FAKAS","",0,300)
$hour = InputBox("Hour for FAKAS","Please input hour:")
$minute = InputBox("Minute for FAKAS","Please input minute:")
WaitRightTime($hour,$minute)
Sleep(2*1000)
For $sample = 1 to 3
   FAKASGoToShortcut($sample)
   FAKASNeedleDownShortcut()
   Sleep(50*1000)
   FAKASNeddleUpShortcut()
Next
```

Code sample 11.4 Script to control FAKAS on a computer assuming that the script in Code sample 11.3 will be used on another computer controlling FACACO.

Save each file on a different computer, and make sure the FACACOFAKASfunctions.au3 file is in the same folder as that of the script on each computer. If you only have one computer, you will need to compile one of the scripts, because SciTE can only run a single script at a time. In order to compile a script, you can press Ctrl + F7, and then save the file as an .exe file. To run the file, you can simply open the folder in which it is saved and double click on it, as it is the case for any executable file.

Before running the scripts, make sure you followed the usual precautions to run FACACO and FAKAS (details in Section 3.5). Also, make sure you provide adequate values for $hour and $minute. For example, if you plan to start the script at 21 : 20, write these values in the codes but then start the scripts at most at 21 : 19, otherwise they will not work as desired. By running the scripts, you should see FACACO and FAKAS operating in synchrony.

Let us examine the codes in detail. They must be familiar to you by now if you have been following the book sequence, and probably the only possibly difficult-to-understand part is the extra sleep commands that are added inside the loop. They substitute the functions that were there in the original script (see Code sample 3.15). The duration of each sleep must be the same of the one or more functions that are being substituted for them. For example, in Code sample 11.3, the command Sleep (11500) substitutes FAKASGoToShortCut and FAKASNeedleDownShortcut, which last, respectively, 11.5 and 5 s (see Code sample 3.14), thus making up 11.5 s (11 500 ms).

11.3 Summary

- A very simple way to synchronize programs being controlled by different computers is by means of time macros.
- Macros are special variables that cannot be modified by the user, and get their value from the Windows operating system.
- Time macros have values referent to the current day, hour, minute, and so on.
- It is essential that computers synchronized using time macros have their time precisely set to be the same, preferably to the same second.

- It is necessary that two scripts are written, one for each computer, because the instruments are controlled by two different computers.
- It is not possible to run two or more scripts at the same time on a single computer. Only one can be run from SciTE, and the others need first to be compiled to an .exe file.

12

Remote Synchronization Using Remote Control Software

Using time macros is a very simple and general solution for the problem of synchronizing instruments being controlled by different computers. However, a much better solution can be achieved if computers can really communicate with each other.

12.1 TeamViewer

Remote control of a computer from another computer has been around for some time, and nowadays there is a very friendly solution: TeamViewer. You can download TeamViewer from https://www.teamviewer.com. It is free for noncommercial use. You can install TeamViewer on several computers and become able to control them from a different computer. This is handy, for example, if you need to go home, but want to check if your instrument is performing well. Also, you can choose to work from your office and not inside a noisy laboratory for at least part of the day. When you run TeamViewer, a window opens showing the screen of the remote computer being controlled. You can control software installed on the other computer seamlessly by means of mouse clicks and keyboard shortcuts. Therefore, it is very easy to use AutoIt to automate software on another computer when TeamViewer is used to connect the computers.

In order to understand how AutoIt can be used together with TeamViewer, let us call a computer a "master" computer, which will start the synchronization, and the other a "follower" computer, which will respond to the initial start (Figure 12.1). The instruments controlled by the master and follower can be synchronized in a very similar way to what was presented in Chapters 3–5 with the only difference that one of the software interfaces will be wrapped in another window. The main limitation that this brings is that it is not possible to access the controls of the software window wrapped by the TeamViewer window. If you use AWI (AutoIt v3 Windows Info), you will see that you cannot get any information about the windows open on the other computer but that appear inside the TeamViewer window; you can only obtain information about the TeamViewer window itself.

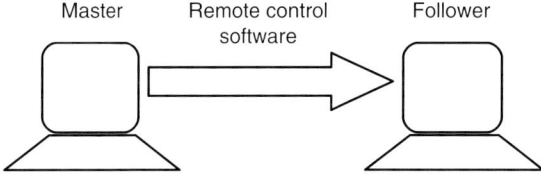

Figure 12.1 A master computer controlling a follower computer by means of remote control software.

12.2 Synchronizing FACACO and FAKAS Using TeamViewer

We can implement a very simple solution to automate FACACO and FAKAS using TeamViewer. In this example, FACACO is installed on the master computer, and FAKAS on the follower computer. We will use controls to automate FACACO, but for FAKAS on the other computer we will rely on mouse clicks, keyboard entries, and pixel monitoring. Therefore, we can use the functions presented in Chapter 5 for FACACO without modification, and those in the end of Chapter 4 for FAKAS, but with some modifications. Therefore, let us begin with the new functions to control FAKAS on the other computer via TeamViewer. However, a few more steps are necessary before that.

First, let us revise the functions in Code sample 4.9. See that their key feature is the use of the function PixelCheckSum to monitor the status area on FAKAS window. In order that the function worked properly, it was essential that FAKAS was at a fixed and known position. Therefore, the first thing we need to ensure is that FAKAS is at a fixed position on the screen of the follower (remote) computer. Then, we need to make sure that the TeamViewer window on the master computer is also at a fixed position. Thus, you could write a very simple script on the follower computer just to ensure that FAKAS will be at the 0,0 pixel. You use the WinMove function for that:

```
opt("WinTitleMatchMode",1)
WinMove("FAKAS","",0,0)
```

Code sample 12.1 Script to ensure FAKAS is at the 0,0 position on the screen of the remote (follower) computer.

Run Code sample 12.1 on the remote computer. Now, FAKAS is at a known position. All the following steps are on the master computer, and the first one is also making sure that the TeamViewer window is at a fixed (e.g., 0,0) position as well. To do so, we need to know the window title, which we can do using AWI. Now, also using AWI, we need to find the position of the corners of the square that we will monitor using the PixelCheckSum function. To do so, make sure FAKAS is visible on the remote computer window. Also, it can be a good idea to set the remote computer scaling parameters to show "original" (you can do this by using the "View" menu on the TeamViewer control bar on the remote computer, see Figure 12.2), so that FAKAS is shown in real size.

Figure 12.2 TeamViewer Toolbar.

Figure 12.3 TeamViewer window wrapping FAKAS window.

Doing that, you can resize the remote computer window to a size that takes only enough space on the master computer screen, the most important aspect being that the status area of FAKAS must be always visible (Figure 12.3).

With these points in mind, let us finally write our functions for the remote control of FAKAS:

```
Func FAKASGoToTeamViewer($input)
    WinActivate("W-140788")
    Mouseclick("left",92,92)
    Send("{BACKSPACE 50}","{DEL 50}")
    Send($input)
    Sleep(1500)
    Send("^g")
    $Psum = PixelChecksum(20,210,150,230)
    While $Psum <> 2312639737
        sleep(20)
        $Psum = PixelChecksum(20,210,150,230)
    WEnd
    Sleep(2000)
EndFunc
Func FAKASNeedleDownTeamViewer()
    WinActivate("W-140788")
    Send("^d")
    $Psum = PixelChecksum(20,210,150,230)
     While $Psum <> 850217397
        sleep(20)
        $Psum = PixelChecksum(20,210,150,230)
    WEnd
EndFunc
Func FAKASNeddleUpTeamViewer()
    WinActivate("W-140788")
    Send("^u")
    Sleep(5000)
EndFunc
```

Code sample 12.2 Functions to remotely monitor and control FAKAS using TeamViewer.

Copy these functions to FACACOFAKASfunctions.au3, and save the file. The first two functions are very similar to those presented in Code sample 4.10, the differences being that WinActivate is used with the TeamViewer window (which on my computer had "W-140788" on its name) and the coordinates for PixelCheckSum are different now. A further difference is that the sleep time before sending the ^g shortcut was extended. This was to ensure that the next action would be done only after the input being passed, which is an extra concern when using TeamViewer as the response times can be delayed on slow connections. As for the third function, it is almost identical to FAKASNeedle-UpShortcut in Code sample 3.14, except for the parameter for WinActivate. Now, we can finally write the script to synchronize FACACO and FAKAS:

```
#include "FACACOFAKASfunctions.au3"
opt("WinTitleMatchMode",1)
WinMove("W-140788","",0,0)
FACACOstartControl()
For $sample = 1 to 3
   Sleep(3*1000)
   FAKASGoToTeamViewer($sample)
   FAKASNeedleDownTeamViewer()
   FACACOmeasureControl($sample)
   FAKASNeddleUpTeamViewer()
   FACACOstatusControl()
Next
```

Code sample 12.3 Script synchronizing FACACO and FAKAS using TeamViewer.

Take the usual precautions for FACACO (see Section 3.5) and run the script. You should see it working as in the previous chapters, but this time one of the programs was really installed on another computer.

The example provided here is probably the simplest type of script you can do using TeamViewer and AutoIt. More sophisticated scripts can be written using the automation features from TeamViewer itself, and not AutoIt. This, however, is beyond the scope of this book. Interested readers can search TeamViewer documentation available online.

A final consideration about using TeamViewer is that it can be unresponsive if the network connection is slow. Although TeamViewer does provide ways to deal with that (e.g., on the remote computer control menu, Figure 12.2, you can choose to optimize speed), there are approaches to the same problem that are far less resource-intensive than TeamViewer. Also, it sometimes happens that antivirus software warns about TeamViewer, and that network administrators may ban its use based on security considerations. Therefore, in the next chapters other solutions to synchronize instruments being controlled by different computers on a network are presented. These solutions are far less demanding regarding network bandwidth, but do involve more scripting work.

12.3 Summary

- TeamViewer is a program that makes it very easy to control a computer from another one, as long as both are connected to the Internet.
- It is not possible to access controls on the remote computer, thus pixel-based monitoring is necessary.
- Despite being of very easy use, TeamViewer heavily relies on network resources, and thus can become slow or unresponsive.

13

Text-Based Remote Synchronization

Synchronization of computers based on time macros or remote control software are the simplest solutions for the problem of synchronizing instruments controlled by different computers. However, there are other still very simple solutions that can be advantageous if given the proper thought. The solution presented in this chapter is light on network resources, and thus can be employed even for slow networks.

The basic condition to enable interactive synchronization between instruments controlled by different computers is that the computers can exchange information between them. Therefore, they need to be connected to a network, such as the Internet. The idea here is to write a script on a computer to send a message (e.g., a single word) from this computer to another. On the other computer, there is a script that is waiting for this message and, when the message arrives, the computer performs a task, and then sends another message back to the first one. The script on the first computer is also waiting for this message, and the cycle can be repeated as many times as needed (Figure 13.1).

Some readers might think that an obvious way to enable such communication between two computers is by sending e-mails from one to another. This is correct, but e-mail can be slow, taking between some tens of seconds and few minutes to reach its destiny. This large and unpredictable time lag prevents the use of e-mail in the context of laboratory automation. Therefore, a better way is to use instant messaging software.

13.1 Choosing Instant Messaging Software

There are numerous options for instant messaging software. I evaluated several of them, and concluded that Trillian is a very good one. Some helpful features of Trillian are: (i) it is free; (ii) it is possible to send messages using controls (see Chapter 5), so that it is not necessary to rely on windows activation, mouse clicks, or keyboard shortcuts; and (iii) it is possible to read the messages sent from another computer also without relying on these actions, this time not using controls, but reading log files.

Before getting to the scripts, let us see some necessary information about Trillian. This program can be downloaded from http://www.trillian.im. You should install it on two different computers so that you can send messages from one to the other, and thus be able to do the exercises in this and the following sections. The version I used for the examples explained here was "5.6 build 5 free."

After downloading and installing Trillian, you need to create an account. It is necessary to create an account per computer. For example, you can create an account for a

13 Text-Based Remote Synchronization

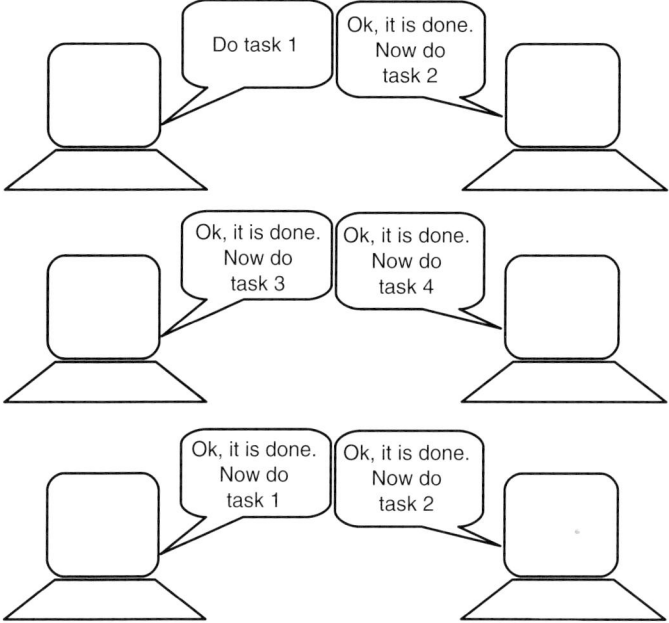

Figure 13.1 Two computers connected to the Internet sending interactive instructions to each other.

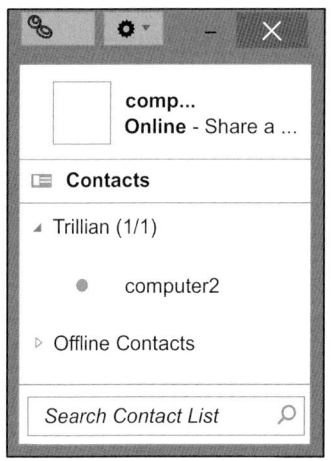

Figure 13.2 Trillian list of contacts, its main interface.

computer with the name computer1, and another for another computer with the name computer2. Once you do that, the interface for Trillian (Figure 13.2) appears. In there, you can add your contacts by clicking on the Trillian button and choosing Add contact. Add computer1 to computer2 and vice versa on the two computers.

Now you can send messages from a computer to the other. Double click on the name of your contact (computer2, in Figure 13.2), and the chat window (Figure 13.3) appears. If you type something and press enter, the message should appear on the other computer, where you can also open a chat window. You can write from the other computer too.

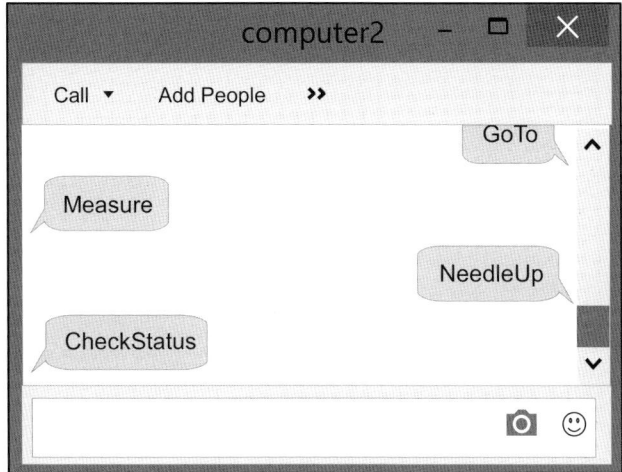

Figure 13.3 Trillian chat window.

13.2 Writing and Reading from Trillian Using AutoIt

It is possible to use AutoIt to automatically send messages to another computer via Trillian. In order to enable that, launch the AutoIt Windows Info utility (AWI) and move the cursor to the window in which messages are being exchanged between the two computers. If the name of the contact you are communicating with is computer2, the window title will have computer2 as its start. Move the cursor to the space where you usually type messages. The control name will appear on AWI, and it should be Trillian Window2. Using this information, it is possible to write the script that sends messages, as below:

```
#include "FACACOFAKASfunctions.au3"
opt("WinTitleMatchMode",1)
ControlSend("computer2","","Trillian Window2","Hello{ENTER}")
```

Code sample 13.1 Sending a message to Trillian using ControlSend.

Run the script, and you should see that the message "Hello" was typed in the message field, and sent to the other Trillian user. If you run this script on the other computer, it should work exactly in the same way, provided you write the window title correctly, since it will be different from the window title on the other computer.

Unfortunately, it is not possible to copy directly the last typed message from the conversation field using AutoIt, which would be useful to use the information passed from one computer to the other. Thus, it is necessary to use another approach to obtain the message, so that AutoIt can work with them. To do so, it is necessary to open the log file of the conversation between the two Trillian users. Such file is created every time a new conversation begins. In order to locate it, click on the Trillian button on the menu bar and select Preferences (Figure 13.4).

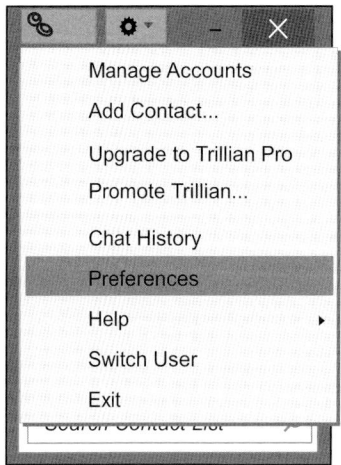

Figure 13.4 Trillian menu with Preferences highlighted.

Then, select Chat History from the window that will open. Click on Chat History, and the window will change, showing the folder where the log files are being saved. You can change this folder, if you wish, but let us not do it now in order to match with the rest of the explanation. If you do not change the storing folder, open this folder, and you should see a folder called ASTRA, and inside this one, another one called Query. This folder contains a text file with the name of your chat partner. Open the file, and you should see the conversation. AutoIt can read the content of text files using specific functions. See the example below:

```
$file = "...\logs\ASTRA\Query\yourpartner.log"
msgbox(0,"Last line of the file",FileReadLine($file,-1))
```

Code sample 13.2 Reading the log file of the chat between two Trillian partners using the command FileReadLine.

The " … " before the rest of the file description needs to be substituted by the location of the log file that you get from Trillian as explained in the previous paragraph. Also, "yourpartner.log" must be replaced by the actual name of the file on your computer. The only new element in the code is the function "FileReadLine." This function receives as arguments the file path (which was attributed to $file) and the line to be read. In this case, the line to be read was line −1 (minus one), which obviously does not exist. In this case, −1 indicates that the last line of the file is read. If you substitute −1 for 1, then the first line will be read, for example.

Run the script, and you should see the last word in the file being presented by the message box. It will appear like this: "computer1: Hello" (supposing that the user name of the sender of the message is computer1). You can use the Code sample 6.5, changing the word being sent, and checking the log file, and you can verify that the message is really being stored.

As explained before, the idea is to send messages from a computer to the other so that the scripts can follow them as instructions. For this, it is convenient that only the final part of the sentence "computer1: Hello," that is, the word "Hello," is isolated from the rest. We learned in Section 8.2 that AutoIt can do that by means of the function StringSplit:

```
$file = "...\logs\ASTRA\Query\yourpartner.log"
$Message = FileReadLine($file,-1)
$MessageArray = StringSplit($Message," ")
msgbox(0,"Last word of the lsat line of the file",$MessageArray[2])
```

Code sample 13.3 Using StringSplit to isolate only the last word of the message stored in the chat log file.

Run the script, and this time you will see that only "Hello" comes in the message box. The function StringSplit takes the contents from $Message and gives them to $MessageArray in the form of an array, which will consist of a set composed of the words in the original string separated between them by empty spaces. The message box shows the second word only, as $MessageArray[2] is being passed as an argument. If, instead, $MessageArray[1] were passed, the first element of the array ("computer1:") would be displayed.

13.3 Synchronizing FACACO and FAKAS Using Trillian

We can use the techniques presented in the previous section to enable the control of FAke Carbon Analyzer Controller (FACACO) and FAKe AutoSampler (FAKAS) in synchrony. The strategy is to send a message to the other computer every time an action by it is needed, and also send back another message every time a subsequent action is needed. First, as in the other chapters, we need to write functions in order to organize the code:

```
Func TrillianReadInstruction($file,$instruction)
    $message = ""
    While $message <> $instruction
        Sleep(10)
        $Content = FileReadLine($file,-1)
        $ArrayContent = StringSplit($Content," ")
        $message = $ArrayContent[2]
    WEnd
EndFunc
```

Code sample 13.4 Function to enable the dynamic reading of an instruction from the log file of the chat on Trillian.

This function utilizes a code very similar to that in Code sample 13.3, but inside a While loop. The purpose of the loop is to keep the script waiting for the instruction until it comes. Let us now see how to integrate FACACO and FAKAS using this approach. As in the chapters dealing with remote synchronization studied so far (Chapters 11 and 12), we need two scripts, one for each computer controlling a different instrument. See the codes below:

```
#include "FACACOFAKASfunctions.au3"
opt("WinTitleMatchMode",2)
$TrillianLogFile = "...\ASTRA\Query\computer2.log"
FACACOstartControl()
For $sample = 1 to 3
   ControlSend("computer2","","Trillian Window2","GoTo{ENTER}")
   TrillianReadInstruction($TrillianLogFile,"Measure")
   FACACOmeasureControl($sample)
   ControlSend("computer2","","Trillian Window2","NeedleUp{ENTER}")
   TrillianReadInstruction($TrillianLogFile,"CheckStatus")
   FACACOstatusControl()
Next
```

Code sample 13.5 Script to control FACACO on a computer assuming that the script in Code sample 13.6 will be used on another computer controlling FAKAS.

```
#include "FACACOFAKASfunctions.au3"
opt("WinTitleMatchMode",2)
$TrillianLogFile = "...\ASTRA\Query\computer1.log"
For $sample = 1 to 3
   TrillianReadInstruction($TrillianLogFile,"GoTo")
   FAKASGoToControl($sample)
   FAKASNeedleDownControl()
   ControlSend("computer1","","Trillian Window2","Measure{ENTER}")
   TrillianReadInstruction($TrillianLogFile,"NeedleUp")
   FAKASNeddleUpControl()
   ControlSend("computer1","","TrillianWindow2","CheckStatus{ENTER}")
Next
```

Code sample 13.6 Script to control FAKAS on a computer assuming that the script in Code sample 13.5 will be used on another computer controlling FACACO.

Before running the scripts, follow the usual precautions to run FACACO and FAKAS (see Section 3.5, for details). FACACO should be installed and opened on the computer running Code sample 13.5, while FAKAS must be installed and opened on the computer running Code sample 13.6. Trillian must be installed on both computers and the user name for the computer controlling FACACO is assumed to be computer1, while for that one controlling FAKAS is assumed to be computer2. If you run one of the scripts, it will get stuck unless you start the other. It does not matter, however, which script is started first. Test these scripts on two different computers with FACACO on one, and FAKAS on the other, and you should see the actions taking place in synchrony. Finally, an important detail: the chat window on Trillian (Figure 13.3) keeps changing its title through the conversation. However, it always keeps the word "computer2" as a part of its title. Then, it was necessary to use opt ("WinTitleMatchMode", 2) instead of 1, as in all previous scripts. Using option 1, the beginning of the window title is used as the reference. With option 2, any portion can be used. Just be careful not to name your scripts with words that could be on the title of the target windows, for example.

Let us examine the codes in detail. In Code sample 13.5, the first real action happens: it is the function FACACOstartControl (see Chapter 5 for explanation about control functions for FACACO and FAKAS). Then, still in this script, the next action is to send the command "GoTo" to the other computer via Trillian, with ControlSend. The other script (Code sample 13.6) has been waiting for this command since this script was initiated. At the same time, the script in Code sample 13.5 keeps waiting for the command "Measure," which is sent also via Trillian after FAKAS gets the "GoTo" command, moves

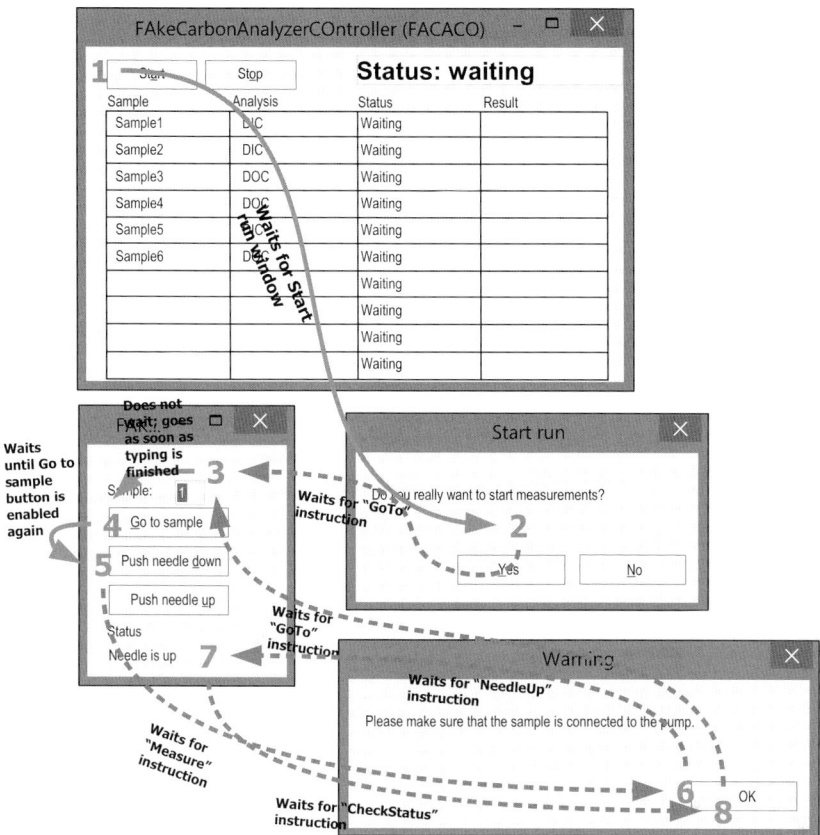

Figure 13.5 Flow of action of the scripts in Code samples 13.5 and 13.6. Large numbers are the steps at which mouse clicks (in this case, not real clicks, but ControlClick) or keyboard inputs (also not real keyboard inputs, but ControlSend or ControlSetText) are done. Continuous arrows link the steps that occur on a single computer. Dashed arrows link the steps that occur between two computers. Texts in black explain the waiting conditions for each step.

the autosampler needle to the correct position, and put this needle down. Then, again in the script controlling FACACO, the function FACACOmeasureControl is called. After that, another instruction is sent to the other computer via Trillian: "NeedleUp," which the other script has been waiting for. Then, this waiting script does one more action, FAKASNeedleUpControl, and sends another message to the other script, "CheckStatus." As in the other instances, this script has been waiting for this instruction, and follows to do the final action, which is FACACOstatusControl. Studying the flow of action (Figure 13.5) for these two scripts may help understand how they work.

13.4 Summary

- Instant messaging enables fast exchange of information between computers connected to the Internet without exhausting network resources.

- The idea is to send simple instructions, like "GoTo" and "Measure" from a computer to the other using an instant messaging program.
- Trillian is an instant messaging program that allows access to controls and stores messages in log files. These two characteristics, and the fact that it is free, make it a very good choice for scripting.
- FileReadLine is a function that reads a line of a text file. If the argument passed as line number is −1, it reads the last line of the file.
- Using StringSplit, the last line of the conversation log file is converted to an array. The element of index [2] in this array corresponds to the desired part of the text, that is, the message that was sent from a computer to another.

14

Remote Synchronization Using IRC

Synchronizing instruments via Trillian was arguably easy, simple, and functional. However, more ambitious scripters may want a solution that does not involve the use of third-party software. An option, in this case, is to use the Internet Relay Chat (IRC) protocol. It is more complicated and demands considerably more knowledge than using instant message software such as Trillian. So, if you prefer not to invest your time into learning this, it is actually not a problem, and you can skip this chapter. However, if for any reason Trillian or other instant messaging programs are banned from your workplace, or if you like the idea of not having to rely on other tools besides AutoIt to do the automation process, then you will be benefited by studying the following contents.

14.1   AutoIt and IRC

Readers born before 1985 may remember that IRC (and its most popular incarnation, mIRC) was the way to chat on the Internet during the last 5 years or so of the twentieth century. New technology as Trillian, Skype, and later Facebook took its place and nowadays IRC is a niche tool used mainly by developers. But it is by no means gone.

It is possible to send messages from a computer to another using the IRC protocol by means of some functions (a library or UDF, see a similar case in Section 6.1.2) available for AutoIt. These functions were developed by McGod (known at the time as Chip) and updated by rcmaehl. They can be downloaded from https://www.autoitscript.com/forum/topic/159285-irc-udf-updated-version-of-chips-irc-udf/.

The most important file on the UDF is "IRC.au3". If you open it using SciTE, you will see a long list of functions. You can explore them by scrolling the file, but also can use the shortcut Alt + Q, which will give you the list of functions in the file (see also Appendix A for a list of useful SciTE features). For most readers of this book, it is probably not easy to write a script from scratch to use the functions listed in IRC.au3. So, I will skip a detailed exploration of its contents and present two simple functions that partially suffice for the purposes of our IRC use, which is to integrate equipment being controlled by different computers (Code sample 14.1):

```
Func _IRCConnectSimple($nick)
    $vSock = _
 _IRCConnect("irc.freenode.net",6667,$nick,0,"RealName","")
    Return $vSock
EndFunc
Func _IRCChannelJoinSimple($sock,$channel)
    _IRCChannelJoin($sock,$channel,"key1")
EndFunc
```

Code sample 14.1 Functions to be added to IRC.au3.

The first function, "_IRCConnectSimple," simply calls the function "_IRCConnect," which is listed in the original IRC.au3 file, with some predefined arguments. Since these arguments are mostly constant, I made the code simpler by including them in a function so that it is not necessary to rewrite them every new script that we need to use _IRCConnect. See that _IRCConnectSimple takes only a single argument, "$nick," while _IRCCOnnect takes five arguments. We will briefly discuss them soon. The other function, "_IRCChannelJoinSimple," also reduces the number of arguments from three to two by calling the original function, "_IRCChannelJoin," and passing it an arbitrary value for one of its arguments.

Let us examine the original IRC.au3 functions mentioned above in order to build the minimal background to be able to use IRC. Let us start with IRCConnect. Its arguments are: server, port, nick, mode, real name, and password. Server is the provider of the IRC service on the Internet. There are several, of which I chose Freenode (irc.freenode.net) because it is currently the most popular. Port is the transmission control protocol (TCP) port, which, for technical reasons, is a number between 6660 and 6669, and a very common one is 6667; there is no need to go deep into these technicalities. Nick is the nickname of the person (or, in this case, the script) entering the chat room. Nick was the only argument that I left open to be edited in _IRCConnectSimple because for each user it will be a different value. Mode is set to zero and later modified in the script; it is not necessary to worry about this. Real name and password can be anything because you almost never need a password. Finally, note that _IRCConnectSimple returns a value, the variable $vSock, which is a socket, which, in computer networking, is one endpoint of a two-way communication link. Therefore, _IRCConnectSimple creates the socket that represents you when connecting to the server.

For _IRCChannelJoin, the arguments are socket, channel, and key. Socket is the value returned by the function _IRCConnectSimple (and consequently by _IRCConnect, see Code sample 14.1). Channel is the medium where communication happens. A channel can be populated by many users, and all of them can exchange messages both publicly and privately; in other words, a channel is a chat room. Finally, key is a password for a channel. I kept the key constant by assigning it a value (key1). However, if you are concerned about security and channel invasion by other IRC users, you can opt by leaving it open, and using _IRCChannelJoin instead of _IRCChannelJoinSimple.

14.2 Monitoring the Connection

Before seeing how we can use the presented functions to exchange messages between computers, let us start using IRC so that it becomes easier to see the scripts in action.

There are many software alternatives that use IRC and enable communication through the Internet. The most popular is mIRC, which can be downloaded from http://www.mirc.com, but Trillian also enables this functionality. I even thought about using mIRC instead of Trillian for Chapter 13, and, in fact, I did. It works in the same way. However, mIRC is not free, and thus I opted for Trillian. Anyway, IRC is very popular and there are several websites that allow IRC connection without any software download. One, for example, is kiwiirc, accessible from https://kiwiirc.com. Let us assume that you use kiwiirc. On the homepage, you choose "try me" and is taken to a login page. There, you can join a channel using a nickname. Choose anything you like for nickname, and, for the channel, choose also a name that suits you, but try something that you believe no one else is using, so you will have a private channel. For example, try #channelXXXX, where XXXX is a big number or a combination of letters and numbers. You do not need to use a password. If you choose right, a new page will open and it will be a chat room. You should see your nick there. You can type in the bottom box that is filled with "Send message ... ". Type anything and press ENTER, and the message should appear in the above box. Repeat this many times and all messages will be displayed there.

Once you are connected to IRC, we can write the necessary scripts. The first script connects to the channel in which you are already logged, and can exchange messages with you. The code presented here was modified from the example provided by rcmaehl and available for download from https://www.autoitscript.com/forum/topic/159285-irc-udf-updated-version-of-chips-irc-udf/, it is the example file:

```
1)  #include "IRC.au3"
2)  $nick = "computer1"
3)  $channel = "#channel45332323"
4)  TCPStartup()
5)  $sock = _IRCConnectSimple($nick)
6)  While 1
7)      $Inputs = _IRCGetMsg($sock)
8)      $ArrayInputs = StringSplit($Inputs, " ")
9)      Switch $ArrayInputs[1]
10)         Case "PING"
11)             _IRCServerPong($Sock, $ArrayInputs[2])
12)         EndSwitch
13)     If $ArrayInputs[0] <= 2 Then ContinueLoop
14)     Switch $ArrayInputs[2]
15)         Case "001"
16)             _IRCChannelJoinSimple($sock, $channel)
17)         Case "JOIN"
18)             $ArrayInputs[3] = StringReplace($ArrayInputs[3],":","")
19)             $ArrayInputs[3] = StringReplace($ArrayInputs[3],@CR,"")
20)             $ArrayInputs[3] = StringReplace($ArrayInputs[3],@LF,"")
21)             _IRCMultiSendMsg($sock, $ArrayInputs[3], "Hello, my friend!")
22)         Case "PRIVMSG"
23)             $sMessage = StringMid($Inputs, StringInStr($Inputs, ":", 0, 2) + 1)
24)             $sMessage = StringReplace(StringReplace($sMessage, @CR, ""), @LF, "")
25)             $aMessage = StringSplit($sMessage, " ")
```

14 Remote Synchronization Using IRC

```
26)             Select
27)                 Case $sMessage = "Hello"
28)                     _IRCMultiSendMsg($sock, $ArrayInputs[3], "Hello, how are you?")
29)                 Case $sMessage = "I am fine, how are you?"
30)                     _IRCMultiSendMsg($sock, $ArrayInputs[3], "Not too bad!")
31)             EndSelect
32)     EndSwitch
33) WEnd
```

Code sample 14.2 Script that enables communication through the channel "#channel45332323" and using the connection arguments listed in Code sample 14.1.

Remember that you should remove the line numbers when copying the code. Make sure you are connected to #channel45332323 and run the script. After a few seconds, you should see the user "computer1" joining the channel, and the message "Hello, my friend!" being displayed in the chat space. After that, there will be silence, unless you type something. If you type the right words, computer1 will reply. For example, type "Hello", and computer1 replies "Hello, how are you?". Type "I am fine, how are you?", to which computer1 replies "Not too bad!". You can leave #channel45332323 and join it again, and computer1 will still be there. The only way to make computer1 leave #channel45332323 is by stopping the script on SciTE. If you want it to join again, just run the script again.

Code sample 14.2 is a long code for the standards of this book, and reflects the higher degree of complexity involved in the script. The first three lines bring nothing new: the necessary library is included, and variables are declared. Then comes the first new function: "TCPStartup." This function starts the use of the TCP protocol, which is necessary for IRC. Then, the variable "$sock" is the socket (explained in Section 14.1) that receives the contents of the function that we defined before, IRCConnectSimple. Finally, there is a big While loop, which is infinite (While 1), and this is why the only way to leave the chat room is by stopping the script on SciTE.

Let us now examine the contents of the While loop. Initially, the variable "$inputs" receives the contents of the function "_IRCGetMsg," which takes $sock as an argument. _IRCGetMsg works by getting the messages that are sent to you from the channel. It is called every run of the loop, its contents can be very variable, and are very important. Most of the rest of the script consists of interpreting what _IRCGetMsg brings. This interpretation consists in analyzing the input, which is done by means of string functions (remember comments in Section 8.2). We already know StringSplit, which splits the string passed as an argument using the next argument (in our case, empty space, or " ") creating an array ("$ArrayInputs," in our code), which has as elements the different words that make up the phrase that was $inputs.

Next, on line 9, $ArrayInputs[1] is checked using the conditional "Switch." Switch works like If ... Then, with the difference that it is easier to use Switch instead of If ... Then when many conditions are possible. The structure is always like this: Switch a certain variable. Case the variable is equal to condition 1, do this. Case it is equal to condition 2, do that. The keyword "Case" is called as many times as needed after Switch. The end of the Switch conditional is marked by "EndSwitch." "Select" can be used in the same way as Switch, see line 26. In our specific case, case the content of $ArrayInputs[1] is "PING," then the function "_IRCChannelJoinSimple" is called. This function simply replies "pong" to the IRC server, which is a way to say that "you are there", otherwise it will close your connection.

Next, $ArrayInputs[0] is checked and, if it is less than 2, the loop continues (function "ContinueLoop"). There is no need to go into these technical aspects, but this line needs to be inserted in the script.

Finally, the largest section of the loop is the check of $ArrayInputs[2]. This section checks this element of the array for some possible inputs. If it is "001," you join the channel. If it is "JOIN," the script processes the incoming information using the function "StringReplace" and also sends a message to the channel using the function "_IRCMultiSendMsg." In this case, the message is "Hello, my friend!". If it is "PRIVMSG," meaning that a message came to you, it again processes the message using string functions, and then starts another conditional (Select), for two possible messages: "Hello" and "I am fine, how are you?", which are each one replied with other messages using the function _IRCMultiSendMsg.

Provided you understood Code sample 14.2, we can now write a similar code so that an automatic conversation takes place in #channel45332323:

```
1) #include "IRC.au3"
2) $nick = "computer2"
3) $channel = "#channel45332323"
4) TCPStartup()
5) $sock = _IRCConnectSimple($nick)
6) While 1
7)     $Inputs = _IRCGetMsg($sock)
8)     $ArrayInputs = StringSplit($Inputs, " ")
9)     Switch $ArrayInputs[1]
10)         Case "PING"
11)             _IRCServerPong($Sock, $ArrayInputs[2])
12)         EndSwitch
13)     If $ArrayInputs[0] <= 2 Then ContinueLoop
14)     Switch $ArrayInputs[2]
15)         Case "001"
16)             _IRCChannelJoinSimple($sock, $channel)
17)         Case "JOIN"
18)             $ArrayInputs[3] = StringReplace($ArrayInputs[3],":","")
19)             $ArrayInputs[3] = StringReplace($ArrayInputs[3],@CR,"")
20)             $ArrayInputs[3] = StringReplace($ArrayInputs[3],@LF,"")
21)         Case "PRIVMSG"
22)             $sMessage = StringMid($Inputs, StringInStr($Inputs, ":", 0, 2) + 1)
23)             $sMessage = StringReplace(StringReplace($sMessage, @CR, ""), @LF, "")
24)             $aMessage = StringSplit($sMessage, " ")
25)             Select
26)                 Case $sMessage = "Hello, my friend!"
27)                     _IRCMultiSendMsg($sock, $ArrayInputs[3], "Hello")
28)                 Case $sMessage = "Hello, how are you?"
29)                     _IRCMultiSendMsg($sock, $ArrayInputs[3], "I am fine, how are you?")
30)                 Case $sMessage = "Not too bad!"
31)                     _IRCMultiSendMsg($sock, $ArrayInputs[3], "Good to hear!")
32)             EndSelect
33)     EndSwitch
34) WEnd
```

Code sample 14.3 Another Script that enables communication through the channel "#channel45332323" and that was designed to "talk" to the script in Code sample 14.2.

Compile this script by pressing Ctrl + F7 on SciTE, so that it can run at the same time as the script in Code sample 14.2. Now, run the.exe file that was created. You should see computer2 logging in #channel45332323. Nothing else will happen unless you log also computer1 to the channel, which is done by running the script in Code sample 6.12. Doing so, you will see a conversation happening between computer1 and computer2. You should also see that, at the end of the conversation, a message box appears.

Code sample 14.3 is very similar to 14.2, the only differences being the contents of the Select conditional and that in the case of JOIN as an input there is no call of the _IRCMultiSendMsg function.

14.3   Synchronizing FACACO and FAKAS

You may have realized that now we have the tools we need to enable synchronization between instruments using IRC. Let us see how it works for FAke Carbon Analyzer Controller (FACACO) and FAKe AutoSampler (FAKAS):

```
1)  #include "IRC.au3"
2)  #include "FACACOFAKASfunctions.au3"
3)  opt("WinTitleMatchMode",1)
4)  $nick = "facaco"
5)  $channel = "#canal1"
6)  $sample = 1
7)  TCPStartup()
8)  $sock = _IRCConnectSimple($nick)
9)  While $sample <4
10)     $Inputs = _IRCGetMsg($sock)
11)     $ArrayInputs = StringSplit($Inputs, " ")
12)     Switch $ArrayInputs[1]
13)         Case "PING"
14)             _IRCServerPong($Sock, $ArrayInputs[2])
15)         EndSwitch
16)     If $ArrayInputs[0] <= 2 Then ContinueLoop
17)     Switch $ArrayInputs[2]
18)         Case "001"
19)             _IRCChannelJoinSimple($sock, $channel)
20)         Case "JOIN"
21)             $ArrayInputs[3] = StringReplace($ArrayInputs[3],":","")
22)             $ArrayInputs[3] = StringReplace($ArrayInputs[3],@CR,"")
23)             $ArrayInputs[3] = StringReplace($ArrayInputs[3],@LF,"")
24)         Case "PRIVMSG"
25)             $sMessage = StringMid($Inputs, StringInStr($Inputs, ":", 0, 2) + 1)
26)             $sMessage = StringReplace(StringReplace($sMessage, @CR, ""), @LF, "")
```

```
27)             $aMessage = StringSplit($sMessage, " ")
28)             Select
29)                 Case $sMessage = "StartFACACO"
30)                     FACACOstartControl()
31)                     _IRCMultiSendMsg($sock, $ArrayInputs[3],
"GoTo")
32)                 Case $sMessage = "Measure"
33)                     FACACOmeasureControl($sample)
34)                     _IRCMultiSendMsg($sock, $ArrayInputs[3],
"NeedleUp")
35)                     FACACOstatusControl()
36)                     $sample = $sample + 1
37)                     _IRCMultiSendMsg($sock, $ArrayInputs[3],
"GoTo")
38)                 EndSelect
39)     EndSwitch
40) WEnd
```

Code sample 14.4 Script controlling FACACO through the channel "#channel45332323," which should be run together with Code sample 14.5 to enable synchrony between FACACO and FAKAS.

```
1) #include "IRC.au3"
2) #include "FACACOFAKASfunctions.au3"
3) opt("WinTitleMatchMode",1)
4) $nick = "fakas"
5) $channel = "#canal1"
6) $sample = 1
7) TCPStartup()
8) $sock = _IRCConnectSimple($nick)
9) While $sample <4
10)     $Inputs = _IRCGetMsg($sock)
11)     $ArrayInputs = StringSplit($Inputs, " ")
12)     Switch $ArrayInputs[1]
13)         Case "PING"
14)             _IRCServerPong($Sock, $ArrayInputs[2])
15)         EndSwitch
16)     If $ArrayInputs[0] <= 2 Then ContinueLoop
17)     Switch $ArrayInputs[2]
18)         Case "001"
19)             _IRCChannelJoinSimple($sock, $channel)
20)         Case "JOIN"
21)             $ArrayInputs[3] =
StringReplace($ArrayInputs[3],":","")
22)             $ArrayInputs[3] =
StringReplace($ArrayInputs[3],@CR,"")
23)             $ArrayInputs[3] =
StringReplace($ArrayInputs[3],@LF,"")
24)             _IRCMultiSendMsg($sock, $ArrayInputs[3],
"StartFACACO")
25)         Case "PRIVMSG"
26)             $sMessage = StringMid($Inputs, StringInStr($Inputs,
":", 0, 2) + 1)
27)             $sMessage = StringReplace(StringReplace($sMessage,
@CR, ""), @LF, "")
```

```
28)                 $aMessage = StringSplit($sMessage, " ")
29)             Select
30)                 Case $sMessage = "GoTo"
31)                     FAKASGoToControl($sample)
32)                     FAKASNeedleDownControl()
33)                     _IRCMultiSendMsg($sock, $ArrayInputs[3], _
"Measure")
34)                 Case $sMessage = "NeedleUp"
35)                     FAKASNeddleUpControl()
36)                     $sample = $sample + 1
37)             EndSelect
38)     EndSwitch
39) WEnd
```

Code sample 14.5 Script controlling FAKAS through the channel "#channel45332323," which should be run together with Code sample 14.4 to enable synchrony between FACACO and FAKAS.

Now, choose Code sample 14.4 or 14.5 to compile, and the other one you can run from SciTE. Remember to have FAKAS and FACACO open, and to take the usual precautions with FACACO (see Section 3.5). Start with Code sample 14.4, which will do nothing until the other one is started. Then, run Code sample 14.5, and you should see FACACO and FAKAS working in synchrony. You should also see the messages appearing in #channel45332323 as the scripts exchange them. Note that watching the messages being exchanged is entirely optional. Note also that the scripts log out automatically at the end of the execution. This happens because the While loops in Code samples 14.4 and 14.5 are not infinite as in Code samples 14.2 and 14.3, but finish when the variable $sample becomes larger than 3.

14.4 Final Considerations

It is important to remember that IRC servers have measures to avoid the proliferation of bots, that is, artificial chatters in chat rooms. This can be a source of problems when running scripts like the one in this chapter. It helps if every time you run the script you change both nick and channel.

Using the technique presented in this chapter, you have a solution to coordinate two computers connected to the Internet. However, it is much more complex than that presented in the previous chapter (Chapter 13), and, actually, less robust due to the problems with IRC server policing bots in chats. Therefore, a strong reason must exist to use this solution instead of that in the previous chapter like, for example, the prohibition of instant messaging software in your workplace.

It is possible to implement even more "elegant" solutions that enable communication between two computers by means of the TCP/IP functionality in AutoIt. However, dealing with network protocols is not straightforward, and can be frustrating. Therefore, I do not recommend such type of approach , unless you have good knowledge on networking.

14.5 Summary

- It is possible to avoid using third-party software to exchange messages between computers by means of directly using IRC protocol, but difficulty to implement this solution is considerably higher.
- IRC stands for Internet Chat Relay, and was a very popular chat platform in the 1990s.
- In order to use IRC protocol with AutoIt, it is necessary to download a library, in which the most important functions are "_IRCCOnnect" and "_IRCChannelJoin."
- I created two new functions that are slight simplifications of the original ones and make coding easier for the purposes outlined in this chapter.
- Although the scripts can be run without any graphical interface, it is good idea to monitor the messages being exchanged using an IRC client.
- Public IRC channels are vulnerable to invasions that may interfere with the proper running of the script.
- IRC channels keep developing techniques of finding robots and banning them. Thus, the scripts presented here are also vulnerable to this problem.
- Overall, there is no real advantage of using the approach presented in this chapter compared to the one based on instant messaging software.

15

Remote Synchronization Using Windows LAN Tools

It is not necessary to have Internet connection in order that computers can synchronize which each other (also excluding the simple case of time macros explained in Chapter 11): it is enough that the computers are connected to a local area network (LAN). It is possible to use the network tools part of the Windows operating system to connect computers to a LAN. Let us see how this can be done so that we can synchronize the software being controlled by different computers. However, be aware that setting up a LAN connection with the Windows tools can be a frustrating exercise. Therefore, Chapter 16 presents an alternative approach in case you fail trying the one presented in this chapter.

15.1 Connecting to a LAN

The hardware necessary to create a LAN can be very simple and inexpensive. For example, you can use a crossover cable and connect two computers, and you have your LAN. Alternatively, you can connect several computers to a router with normal network cables or wirelessly, and you will also have a LAN. Crossover cables, network cables, and routers are available at most electronics stores.

Let us see here the simplest case, that is, you have two computers connected to each other using a crossover cable. If this is the case, click on the start button and type "ncpa.cpl" (if you are using Windows XP, you need to choose "run" first), then press enter. This is the easiest and more direct way to access the existing networks that can be connected to your computer (Figure 15.1).

If you are connected via a crossover cable to another computer, the "Ethernet" or "local area connection" (this depends on the computer) icon will show "Unidentified network." If you are not connected, it will show "Network cable unplugged." Once you verify you are connected, you need to know the IP addresses of all computers in the network (in this case, only two). You can do this using the ipconfig on the command prompt. Go to the start button, and type command prompt (if you use Windows XP, choose run first). The command prompt (Figure 15.2) allows you to type commands that can perform several useful tasks.

Command prompt is one of the easiest ways to find the IP address of your computer, by means of the ipconfig command. Type it and then press enter. You should see a list with several IP addresses depending on how many networks your computer is connected to (Figure 15.3).

Practical Laboratory Automation: Made Easy with AutoIt, First Edition. Matheus C. Carvalho.
© 2017 Wiley-VCH Verlag GmbH & Co. KGaA. Published 2017 by Wiley-VCH Verlag GmbH & Co. KGaA.

▶ Control Panel ▶ Network and Internet ▶ Network Connections ▶

Figure 15.1 Location accessed using the command ncpa.cpl on the start button for Windows 8.1 which shows a list of Networks available for a computer found using ncpa.cpl.

Figure 15.2 Command prompt on Windows 8.1.

```
C:\>ipconfig

Windows IP Configuration

Ethernet adapter Local Area Connection:

   Media State . . . . . . . . . . . : Media disconnected
   Connection-specific DNS Suffix  . :

Wireless LAN adapter Local Area Connection* 3:

   Media State . . . . . . . . . . . : Media disconnected
   Connection-specific DNS Suffix  . :

Ethernet adapter Bluetooth Network Connection:

   Media State . . . . . . . . . . . : Media disconnected
   Connection-specific DNS Suffix  . :

Wireless LAN adapter Wi-Fi:

   Connection-specific DNS Suffix  . : acad.ssu.ad
   Link-Local IPv6 Address . . . . . : fe80::4013:817b:561f:9dw9%3
   IPv4 Address. . . . . . . . . . . : 10.130.102.225
   Subnet Mask . . . . . . . . . . . : 255.255.252.0
   Default Gateway . . . . . . . . . : 10.130.100.1

Ethernet adapter Ethernet:

   Connection-specific DNS Suffix  . :
   Link-Local IPv6 Address . . . . . : fe80::8d14:e564:bc47:77d2%3
   IPv4 Address. . . . . . . . . . . : 169.254.119.211
   Subnet Mask . . . . . . . . . . . : 255.255.0.0
   Default Gateway . . . . . . . . . :

Tunnel adapter isatap.acad.ssu.ad:

   Media State . . . . . . . . . . . : Media disconnected
   Connection-specific DNS Suffix  . : acad.ssu.ad

Tunnel adapter isatap.{D1A53FFF-5340-4888-810D-EEF088D900B4}:

   Media State . . . . . . . . . . . : Media disconnected
   Connection-specific DNS Suffix  . :

C:\>
```

Figure 15.3 Typical result for the ipconfig command. In this case, the IP address that we need is the IPv4 for the Ethernet adapter, which is 169.254.119.211. This number will be probably different on your machine.

▸ Network ▸ 169.254.105.223

Figure 15.4 Path of computer 166.254.105.223 accessed from computer 169.254.119.211 via LAN.

The IP address that you need to find, in this case, is the one for the Ethernet connection. Therefore, in the list given by ipconfig (Figure 15.3), you need to find Ethernet adapter. The number you are after is the IPv4, which in the example in Figure 15.3 is 169.254.119.211. On the other computer connected through the crossover cable, the number will be different; in this case, it was 169.254.105.223.

Let us have a closer look at IP addresses. IP stands for "Internet Protocol." It is a number that identifies a computer in a network. The standard configuration on Windows is that this number is automatically generated. You can choose to change it, if you want, but it is not necessary in our example here. There are two types of IP addresses: IPv4 and IPv6. IPv4 is the older version of IP, and IPv6 the newer. IPv4 still is the norm and suffices for our purposes. However, it is expected that in some years IPv6 will replace it. For now, let us stick to IPv4. ipconfig listed several IP addresses for the computer in Figure 15.3. This is because the IP address of a computer is different for each network to which it is connected. This is why it was important to know the exact network (in this case, the Ethernet adapter) to which the other computer was connected.

Once we know the IP addresses of the computers in our network, we can access files in both of them. Let us see how. Again, go to the start button, and then type the IP address of the other computer connected to the LAN. In our case, it was 169.254.105.223, and thus you need to type \\169.254.105.223. Do not forget the \\. Doing so, you would access the path shown in Figure 15.4.

Usually, it is when you try to get the window in Figure 15.4 that problems occur, and sometimes it is not possible to establish a connection between computers. If you are unable to see a window like that in Figure 15.4, but instead you see a message saying that you need a password or one saying that it is impossible to connect to the other computer, you will need to play with your network setting on at least one of the computers in the network. It is not the purpose of this book to explore all the potential problems in this specific case, so you need to try to find out a solution for your problem in other sources. Unfortunately, my experience shows that it can be difficult to find such solutions. Check the next chapter in case you cannot keep following the approach here.

15.2 Creating a Shared Folder

Supposing that things go well and you can see a window similar to that in Figure 15.4 on your computer, the next step is to create a shared folder in which we will have a file that will be used to exchange instructions between the computers. In one of your computers, go to the root folder (C:\) and create a folder named "Shared" (if you already have a folder with this name, you will need to use another name, and remember to modify the scripts to adapt to your case). Once you create this folder, open it, and right-click on its window so that you see the option "Share with," which opened has the option "Specific people … " (Figure 15.5).

Click on Specific people (Figure 15.5) and another window will open, where you can choose with whom you want to share the contents of the folder (Figure 15.6).

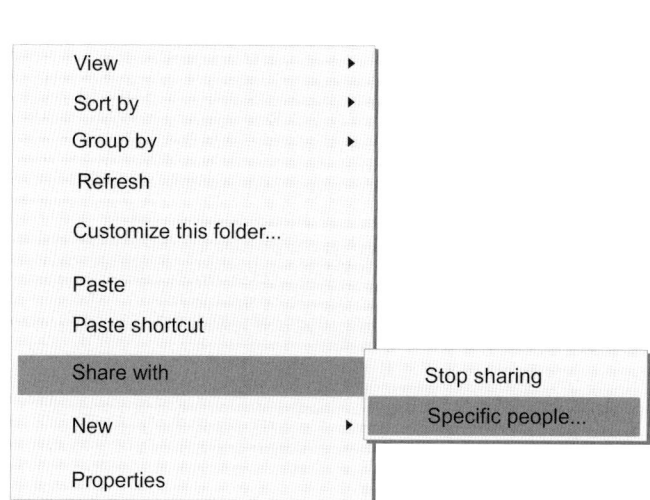

Figure 15.5 Menu opened using the right button to share a folder.

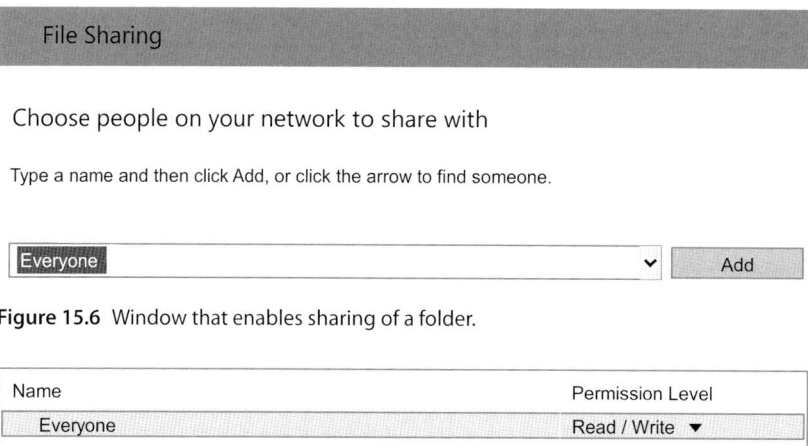

Figure 15.6 Window that enables sharing of a folder.

Figure 15.7 Message showing that the folder that now can be shared with "Everyone."

In the window shown in Figure 15.6 (it can be a little different depending on the Windows version on your computer), choose "Everyone" in the drop-down menu near the "Add" button. By doing so, now Everyone will be listed among the people who can access this folder (Figure 15.7).

Before pressing the "Share" button, which would make the folder available for all computers in the LAN, there is a final step, which is to make the folder writable. You do that by clicking on the "Read" just below "Permission Level." A menu will appear with "Read," "Read/Write," and "Remove" as options. Choose "Read/Write" (Figure 15.7) and then finally press the Share button. Make sure you do not skip this step, otherwise the

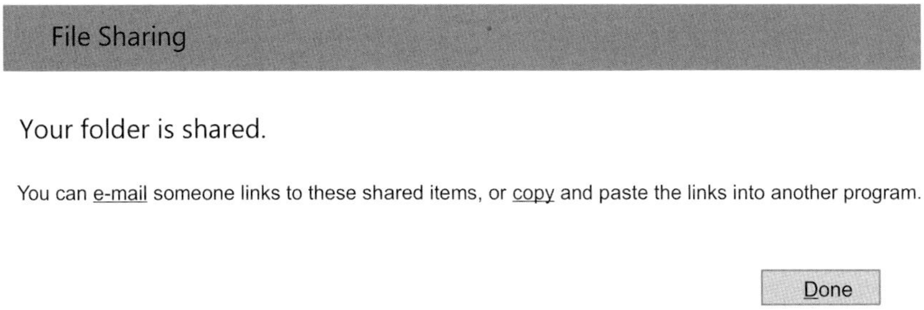

File Sharing

Your folder is shared.

You can e-mail someone links to these shared items, or copy and paste the links into another program.

Done

Figure 15.8 Message confirming that now the folder is shared in the LAN.

Figure 15.9 Window showing computer 166.254.105.223 accessed from computer 169.254.119.211 via LAN, this time listing the Shared folder.

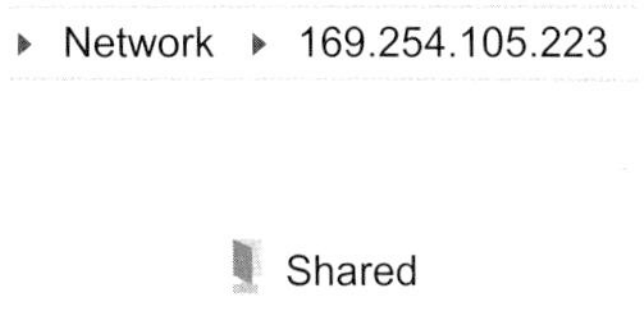

▸ Network ▸ 169.254.105.223

Shared

scripts will not work. Once you press share, another window will confirm that now the folder is shared in the LAN (Figure 15.8).

Now, go back to the window showing the contents of 166.254.105.223. If you refresh its contents, now you should see the Shared folder listed there (Figure 15.9).

Now you can open the folder and list its contents. It should be empty now. Note that this folder only needs to be created on one of the computers. Here, it was created on 166.254.105.223 and the window in Figure 15.9 is what is seen from 169.254.119.211.

15.3 Synchronizing FACACO and FAKAS

Once we have a shared folder, we can proceed to write our scripts to synchronize (FAke Carbon Analyzer COntroller) FACACO and FAKAS, each on a different computer. The strategy will be to use a .txt file in the Shared folder as a repository of instructions exchanged between the computers. The first step is to use a text editor like Notepad to create a .txt file in the folder; name it "instructions.txt". Instructions will be passed from the scripts using the function "FileWrite". Before that, let us write a function that waits for an instruction, like the one in Chapter 13 (Code sample 13.4) for Trillian:

```
Func LANReadInstruction($file,$instruction)
    $message = ""
    While $message <> $instruction
        Sleep(10)
        $message =FileReadLine($file,-1)
    WEnd
EndFunc
```

Code sample 15.1 Function to enable the dynamic reading of an instruction sent to instructions.txt in the shared folder.

Type this function in FACACOFAKASfunctions.au3 and save the file. It is a very simple function, and works in the same way as the one presented in Code sample 13.4.

As you may recall (Section 13.2), FileReadLine with a −1 as the second argument means that the last line of the file will be read. Therefore, the file will be constructed by adding new lines with the instructions at each line. Let us now see the codes to automate FACACO and FAKAS:

```
#include "FACACOFAKASfunctions.au3"
$SharedFile = "\\169.254.105.223\shared\instructions.txt"
FACACOstartControl()
For $sample = 1 to 3
   FileWrite($SharedFile,"GoTo"&@CRLF)
   LANReadInstruction($SharedFile,"Measure")
   FACACOmeasureControl($sample)
   FileWrite($SharedFile,"NeedleUp"&@CRLF)
   LANReadInstruction($SharedFile,"CheckStatus")
   FACACOstatusControl()
Next
```

Code sample 15.2 Script that must run on the computer controlling FACACO in order to synchronize it with the other computer controlling FAKAS the LAN configurations on Windows.

```
#include "FACACOFAKASfunctions.au3"
$SharedFile = "C:\shared\instructions.txt"
For $sample = 1 to 3
   LANReadInstruction($SharedFile,"GoTo")
   FAKASGoToControl($sample)
   FAKASNeedleDownControl()
   FileWrite($SharedFile,"Measure"&@CRLF)
   BingoReadInstruction($SharedFile,"NeedleUp")
   FAKASNeddleUpControl()
   FileWrite($SharedFile,"CheckStatus"&@CRLF)
Next
```

Code sample 15.3 Script that must run on the computer controlling FAKAS in order to synchronize it with the other computer controlling FACACO the LAN configurations on Windows.

Similarly to the procedure in Chapter 13, you will need to save the first file on one computer, which will run FACACO, and the second one on the other computer, which will run FAKAS. Take the usual precautions with FACACO (Section 3.5). Run the script on the computer controlling FAKAS, and then on the one running FACACO. As in all previous chapters dealing with synchronization, once more you should see both programs working perfectly in synchrony.

As mentioned earlier, the instructions were passed using the function FileWrite. Each instruction consisted of a single word, and was followed by a macro that means line break (@CRLF). This ensured that each line of the file had only one instruction.

15.4 Summary

- The Windows operating system offers resources to connect computers using a LAN.
- Computers can connect to a LAN using LAN cables or via Wi-Fi.
- The easiest way to find your LAN configurations is using the command ncpa.cpl.
- Many times, a signal that your computer is connected to a LAN is indicated by a status of "Unidentified network."

15.4 Summary

- Once you are connected to a LAN, you need to know your and the other computer's IP addresses for that specific LAN.
- The IP address of your computer can be known by launching the command prompt and using the command ipconfig.
- Ipconfig can return several IP addresses. Be sure to identify the correct one for the LAN that is being used for synchronization between computers.
- Once you know the IP address of the other computer in the LAN, you should be able to access it by typing its IP address on the start menu.
- Sometimes, there are problems when trying to access a computer in this way. If this happens, you either need professional support or should try the approach explained in the next chapter.
- If you can access the other computer on the LAN, you need to create a new shared folder in it.
- The shared folder will contain a file that will be used to exchange text between the two computers.
- Once this is enabled, an approach very similar to that of Chapter 13 can be used to synchronize computers.

16

Remote Synchronization Using Third-Party LAN Software

As mentioned in the previous chapter, it can be frustrating to try to use the standard Windows setup tools for LAN connection. This is because network setup is subject to many different configurations, and often there are restrictions on what a computer can achieve. These restrictions are commonly implemented by the IT personnel in an institution and cannot be overridden to comply with the local policy. Therefore, a third-party program is presented, which makes the task much easier.

16.1 Connecting to a LAN Using Bingo's Chat

I tried some different options and found that a very good program to connect computers via LAN is Bingo's Chat, which can be downloaded for free from http://bingosoft.info/downloads/bingochat/. Note that the site is not in English. However, the program itself is in English. Select the most recent version (at the time of writing this, it was 1.6.10), install it following the standard instructions, and you should be ready to go. You will need to install the program in the two (or more) computers that are to be connected. The program has many fancy interfaces, so here only the elements with some relevance have been shown for our purpose. For example, when starting the program, your computer is listed at the right-hand side of the window (Figure 16.1), together with its IP address.

In Figure 16.1, Bingo's Chat is shown with only a single user. Most of the times, if you install the program on two or more different computers that are connected to a LAN, they will appear automatically in the "Users" list at the right-hand side of the interface. If, however, you start the program and cannot see each other, it is possible that Bingo chose a different network to connect. In order to change the network to which you are connected, and find the other computers, you go to the Options menu, then choose Client–Client, then Network interface (Figure 16.2).

Then, a new window will appear (Figure 16.3), where you can choose the IP address that you want. As we saw in the previous chapter, it is important to choose the correct IP address because a computer can have many of these addresses, one for each network to which it is connected. Once you choose the correct one (here, it is 169.254.119.211), you can click on Ok, and you should start being able to see the other computers on the LAN. If not, restart Bingo on all computers in the LAN, and then it should work.

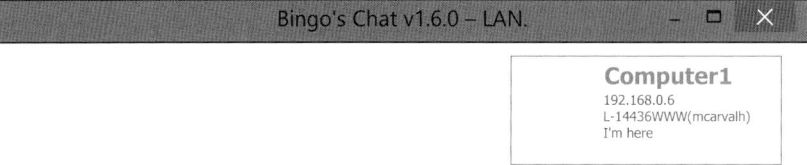

Figure 16.1 Computer listed along with its IP address on Bingo's Chat interface.

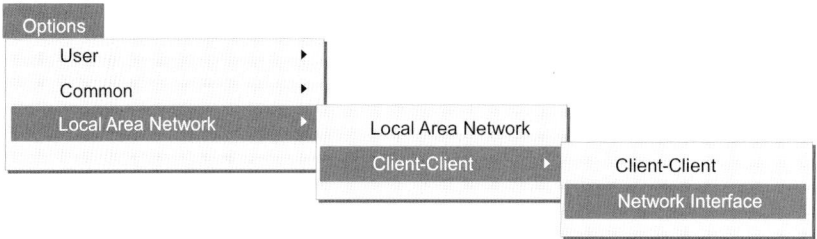

Figure 16.2 Opening the menu to change the network being accessed by Bingo's Chat.

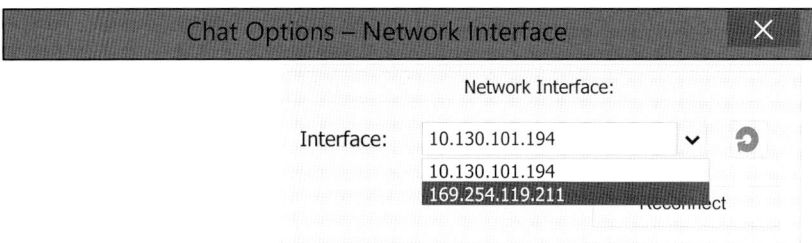

Figure 16.3 Choosing the correct IP address for a LAN connection using Bingo's Chat.

16.2 Automated Communication Using Bingo's Chat

Once you have two computers on the Bingo interface (Figure 16.4), you can finally send messages to each other. You can use either the "Chat" tab or the one in which the name of the other computer is written. We will use here that with the name of the other computer.

Using the communication tabs, we can send messages to other computers. Using AWI, we find the name of the control, which is "TsMemo1." Therefore, we can use ControlSend and automatically send messages through the interface. The next step is to read the sent messages. As for Trillian (Chapter 13), we will use the log file for this. The first step is to find the log file, which can be done by opening the File menu, then choosing View Logs (Figure 16.5).

By choosing that option, a new window opens (Figure 16.6) where you can locate the log file for your conversation. It will be under the name of the chat partner (in this case, I was using Computer1, so I found it under Computer2), and will be named with the date of the conversation, in this case, 23.11.2015.htm. You can use the search feature of Windows, but a very good tool to find files on Windows is Everything, which can be

16.2 Automated Communication Using Bingo's Chat

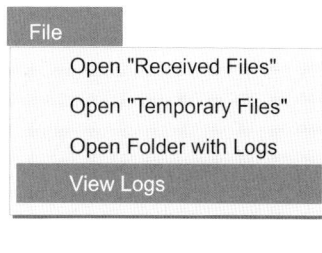

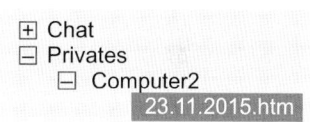

Figure 16.4 Elements of Bingo's Chat interface with two computers connected.

Figure 16.5 Finding the log files of conversations using Bingo's Chat.

Figure 16.6 Log files of Bingo's Chat.

downloaded from http://http://www.voidtools.com. Using this tool, you can easily find the file and get its full path name.

The log file for Bingo's Chat is a .htm file, and not a text file, as it was for Trillian (Section 13.2). This brings some extra difficulties when trying to retrieve the useful information for using in the scripts. If you open the log file, it will probably open on your Internet browser (.htm is the standard format for web pages). You should see the conversation exchanged between the two different computers (Figure 16.7).

Looking at Figure 16.7, you could think that an approach like the one for the log file of Trillian would work, since you only need to get the last word of the last line and thus know the last message sent in the chat. However, it is not like that. Web pages have a source code, in the same way your script does. In order to get the instruction from a web page and use in a script, you need to use its source code. There are several ways to check the source code for a web page. Your web browser may have this capability, or you can choose Notepad to open the file, and the source code will appear.

-=•(20-51-55-Mon)•=- **Computer2** : Hi
-=•(20-52-23-Mon)•=- **Computer1** : Hi

Figure 16.7 Portion of a Bingo's Chat log file open on a web browser.

```
<html><head><title>Bingo Chat Logs</title><meta HTTP-EQUIV="Content-
Type" CONTENT="text/html; charset=windows-1251"><style
type="text/css">A:link {   COLOR: #0000FF; TEXT-DECORATION:
none}A:visited {COLOR: #0000FF; TEXT-DECORATION: none}A:active
{COLOR: #0000FF}A:hover {FONT-WEIGHT: normal; COLOR: #00FF00; TEXT-
DECORATION: underline}</style></head><body bgcolor="FFFFFF"
LEFTMARGIN="0" TOPMARGIN="0" MARGINWIDTH="0" MARGINHEIGHT="0"><font
style="font-size:16pt" face="Arial"   color=#808080><br>-=•(13-41-
05·Mon)•=- </font><font style="font-size:10pt" face="Arial"
color=#808080> Measure</font><font style="font-size:10pt"
face="Arial" color=#808080><br>-=•(13-50-26·Mon)•=- </font><font
style="font-size:10pt" face="Arial" color=#808080> :
Teste</font><font style="font-size:10pt" face="Arial"
color=#808080><br>-=•(13-50-30·Mon)•=- </font><font style="font-
size:10pt" face="Arial" color=#808080> Teste2</font><font
style="font-size:10pt" face="Arial" color=#808080><br>-=•(13-50-
43·Mon)•=- </font><font style="font-size:10pt" face="Arial"
color=#808080> : GoTo</font><font style="font-size:10pt" face="Arial"
color=#808080><br>-=•(13-50-53·Mon)•=- </font><font style="font-
size:10pt" face="Arial" color=#808080> Measure</font><font
style="font-size:10pt" face="Arial" color=#808080><br>-=•(13-50-
58·Mon)•=- </font><font style="font-size:10pt" face="Arial"
color=#808080> : NeedleUp</font><font style="font-size:10pt"
face="Arial" color=#808080><br>-=•(13-51-03·Mon)•=- </font><font
style="font-size:10pt" face="Arial" color=#808080>
CheckStatus</font><font style="font-size:10pt" face="Arial"
color=#808080><br>-=•(13-51-09·Mon)•=- </font><font style="font-
size:10pt" face="Arial" color=#808080> : GoTo</font><font
style="font-size:10pt" face="Arial" color=#808080><br>-=•(13-51-
19·Mon)•=- </font><font style="font-size:10pt" face="Arial"
color=#808080> Measure</font><font style="font-size:10pt"
face="Arial" color=#808080><br>-=•(13-51-24·Mon)•=- </font><font
style="font-size:10pt" face="Arial" color=#808080> :
NeedleUp</font><font style="font-size:10pt" face="Arial"
color=#808080><br>-=•(13-51-28·Mon)•=- </font><font style="font-
size:10pt" face="Arial" color=#808080> CheckStatus</font><font
style="font-size:10pt" face="Arial" color=#808080><br>-=•(13-51-
55·Mon)•=- </font><font style="font-size:10pt" face="Arial"
color=#808080> : GoTo</font><font style="font-size:10pt" face="Arial"
color=#808080><br>-=•(13-52-05·Mon)•=- </font><font style="font-
size:10pt" face="Arial" color=#808080> Measure</font><font
style="font-size:10pt" face="Arial" color=#808080><br>-=•(13-52-
10·Mon)•=- </font><font style="font-size:10pt" face="Arial"
color=#808080> : NeedleUp</font><font style="font-size:10pt"
face="Arial" color=#808080><br>-=•(13-52-14·Mon)•=- </font><font
style="font-size:10pt" face="Arial" color=#808080> CheckStatus</font>
```

Code sample 16.1 HTML source code for a log file of Bingo's Chat. This is NOT an AutoIt script.

You are probably confused with Code sample 16.1. It consists of HTML, the code used to publish web pages. Fortunately, you do not need to understand it for the purposes of this book. The only things you need to know are: (i) the last instruction comes at the end of the single line that makes up the file and (ii) after this instruction, comes the string </font>. Using this information, we can write the code to get the instruction from the log file:

```
Func BingoReadInstruction($file,$instruction)
   $message = ""
   While $message <> $instruction
      Sleep(10)
      $Content =FileRead($file)
      $ArrayContent = StringSplit($Content," ")
      $ArrayLength = Ubound($ArrayContent)
      $FinalWord = $ArrayContent[$Arraylength-1]
      $FinalSubword = StringSplit($FinalWord,"<")
      $message = $FinalSubword[1]
   WEnd
EndFunc
```

Code sample 16.2 Function that reads the instruction from Bingo's Chat log file.

This function works like the one in Code sample 13.2: it reads the last useful word in the log file and compares to the expected instruction. While the words are not the same, it keeps repeating the procedure every 10 ms. "FileRead" passes to "$Content," the whole content of the file. Then, $Content, which is a string, is split into many different elements using the function StringSplit, creating the array "$ArrayContent" (for more details about strings and arrays, see Chapter 8). The size of the array changes every time a new instruction is given and stored in the log file. Therefore, we cannot use a fixed number to get the last element of the array. The solution is to use the function "Ubound," which gives the length of the array to the variable "$ArrayLength" (see Section 8.2). Once we know the size of the array, we can isolate its final element, which is $ArrayContent[$ArrayLenght – 1]. It is necessary to subtract 1 due to particularities of the AutoIt language (Section 8.1). If you look at Code sample 16.1, you will see that the final word, as separated by spaces, which we determined using StringSplit, is "Check-Status." Thus, we use StringSplit again, but this time we choose the separator to be "<", and not " ". This way, its first element will be "CheckStatus," which we can compare with the variable "$message" to decide whether the correct instruction was given or not. Save this code in the FACACOFAKASfunctions.au3 file, as usual.

16.3 Synchronizing FACACO and FAKAS

Now we have all the tools to write the scripts that synchronize (Fake Carbon Analyzer Controller) FACACO and FAKAS being controlled by different computers on a LAN and interfaced by Bingo's Chat:

```
#include "FACACOFAKASfunctions.au3"
opt("WinTitleMatchMode",1)
$BingoLogFile = "file path, check on your computer"
FACACOstartControl()
For $sample = 1 to 3
   ControlSend("Bingo","","TsMemo1","GoTo{ENTER}")
   BingoReadInstruction($BingoLogFile,"Measure")
   FACACOmeasureControl($sample)
   ControlSend("Bingo","","TsMemo1","NeedleUp{ENTER}")
   BingoReadInstruction($BingoLogFile,"CheckStatus")
   FACACOstatusControl()
Next
```

Code sample 16.3 Script that controls FACACO on a computer connected to another one controlling FAKAS, and running Code sample 16.4, via LAN and interfaced with Bingo's Chat.

```
#include "FACACOFAKASfunctions.au3"
opt("WinTitleMatchMode",1)
$BingoLogFile = "file path, check on your computer"
For $sample = 1 to 3
   BingoReadInstruction($BingoLogFile,"GoTo")
   FAKASGoToControl($sample)
   FAKASNeedleDownControl()
   ControlSend("Bingo","","TsMemo1","Measure{ENTER}")
   BingoReadInstruction($BingoLogFile,"NeedleUp")
   FAKASNeddleUpControl()
   ControlSend("Bingo","","TsMemo1","CheckStatus{ENTER}")
Next
```

Code sample 16.4 Script that controls FAKAS on a computer connected to another one controlling FACACO, and running Code sample 16.3, via LAN and interfaced with Bingo's Chat.

Take the usual precautions (see Section 3.5) and run the script. As usual, FACACO and FAKAS should work in synchrony. Remember to write the correct path name for your file on both scripts. Note that they will be different on each computer.

A limitation of the approach presented here is that the log file has a date as its name. Therefore, if the script is run overnight, it will stop working properly after midnight. You can try to write another script to work around this, or can try a different program. Trillian and Teamviewer both have support for LAN. I tried other programs that do the same job, but their controls were invisible to AWI. However, this does not mean that a better program does not exist. The great advantage of AutoIt is that you do not need to rely on any specific piece of software to do the desired job. If you find a better solution than Bingo's Chat, you can certainly easily adapt the technique presented here.

16.4   Summary

- Sometimes, the Window LAN configurations do not allow communication between computers.
- In this case, a solution is to rely on programs that overcome such problems, such as Bingo's Chat.
- As for when connecting to a LAN using Windows configurations, it is important to know the correct IP addresses involved.
- Once this is established, an approach very similar to that for Trillian in Chapter 13 can be used to synchronize programs on different computers.
- The chat log file for Bingo's Chat is a .htm file, and not a .txt file. This brings complications when getting the correct instruction, but these can be dealt with by analyzing the file code and using StringSplit.

17

Interacting with Devices via COM Ports

So far throughout the book, we have learned how AutoIt can control the programs that control different instruments (Figure 1.2). However, AutoIt can also be used to directly control instruments. An essential aspect of controlling instruments with a computer is the ability to exchange data between computer and instrument. This is made possible through communication interfaces, like serial, parallel, USB, and LAN ports, among many other ways. Specialized technical knowledge is necessary to implement any of these technologies. It is not the purpose of this book to explore this subject; in fact, the idea of the book so far has been to avoid it altogether. However, AutoIt does have the ability to deal with communication between computer and instruments, and this can be useful in some instances.

In today's laboratories, one of the most common ways of connecting instruments to a computer is using USB or serial ports, which are based on the RS-232 protocol. This chapter introduces how AutoIt can send commands to instruments using this technology.

17.1 Serial Communication Protocols

Before going to the practical aspects of using communication ports with AutoIt, here the absolute minimum will be presented that needs to be known about communication protocols in order to use AutoIt to control a device using USB or serial ports (from now on, USB and serial will be collectively called serial, to simplify; USB, by the way, means Universal SERIAL Bus).

Serial communication is based on the RS 232 protocol. This protocol was developed to standardize communication between electronic devices. Many devices used in the laboratory are controlled using a variation of this technology. In most cases, the relevant details are not made public, so that the manufacturer can obligate the user to buy the device together with the other one that complements it from the same manufacturer. Throughout this book, we learned how to overcome this limitation using scripting for the cases in which the instrument has a graphical user interface for the Windows operating system (the most common case).

There are cases, however, in which a device is developed with the purpose of being used together with others of several different manufacturers, and for these the

Practical Laboratory Automation: Made Easy with AutoIt, First Edition. Matheus C. Carvalho.
© 2017 Wiley-VCH Verlag GmbH & Co. KGaA. Published 2017 by Wiley-VCH Verlag GmbH & Co. KGaA.

manufacturer makes the communication protocols available. This information, for serial communication, consists of the following parameters:

1) Baud rate, which is the speed of communication in bits per second. A common value is 9600.
2) Number of bits, which is the size of each unit of data being transmitted. A typical value is 8.
3) Parity bits, which are used for error checking. Many times they are not used.
4) Stop bits, which are used to mark the end of a data unit. A typical value is 1.
5) Flow control, which is used to adjust the flow of data so that it can be slowed and no information is lost. In many cases, there is no flow control.
6) Handshaking, which is used to organize the exchanged information. Not used in many cases.

Each instrument designed with the purpose of being connected to a computer using COM ports will probably have information available in its manual about the parameters listed above. For the purposes outlined in this book, there is no need to explore more about the details of serial communication. Interested readers should seek information in other sources, as, for example, the one listed at the end of this chapter.

17.2 AutoIt and COM Ports

There is a library (normally called UDF, or user-defined functions) for AutoIt, which allows the control of instruments through COM ports. It can be downloaded from http://www.mosaiccgl.co.uk/AutoItDownloads/confirm.php?get=COMMGvv2.zip, and was developed by Martin.

Download the files and copy them to the folder where you write your scripts. Then, open the file named "CommMG.au3." This file contains the functions that can be used to control devices through serial ports. It also contains detailed technical details about each function. For us, the most useful functions are: (i) "_CommSetPort," which sets the communication port to be accessed; (ii) "_CommSendString," which sends commands to the devices connected to the communication port; and (iii) "_CommGetString," which reads information from the device being controlled. Using these three functions, it is possible to control and monitor devices connected via a serial port to a computer. If it is being difficult for you to locate the functions in the file, on SciTE press Alt + Q, this will open a window with all the functions in the code (see also Appendix A for more details). If you click on the function name in the list, SciTE will bring the cursor to the selected function in the code.

Before going to examples, some other functions will be created and added to the commMG.au3 library. The first one is "_CommSendStandard":

```
Func _CommSendStandard($port,$error,$command)
   _CommSetPort($port, $error, 9600, 8, 0, 1, 2)
   if $error <> '' Then
      MsgBox(262144, 'Port Error = ', $error)
   EndIf
   _CommSendString($command & @CR)
EndFunc
```

Code sample 17.1 _CommSendStandard, a new function for sending information to a serial port.

Add this function to the file CommMG.au3 so that it can be called by the subsequent scripts when including the file in the code. _CommSendStandard does three things: (i) it calls the function _CommSetPort, which sets the port to be used with the default parameters (baud rate = 9600, number of bits = 8, parity bits = 0, stop bits = 1, and no flow control); (ii) it checks for errors, like, for example, wrong port number or device connected to the port number already being used by another software; and (iii) it sends the command to the device connected to the port using the _CommSendString function.

The other new function is "_CommReadString," which is developed over the original _CommGetString. The usage of the _CommGetString function is not straightforward: instead of working like, for example, an input box, in which the typed words are recorded and can be used, this function keeps reading constantly the information that comes from the communication port, and thus needs to be called constantly in order that the information is collected. In addition, the information obtained this way needs to be pooled in a block, otherwise only part of it is used. For the examples listed below a different type of function is more useful, one that stores only the useful data that _CommGetString collects for a period of time. The code for the function goes below:

```
Func _CommReadString($TimeSpan)
   Local $Reading
   $Start = TimerInit()
   $End = 0
   $ReadStep = 0
   While $End < $TimeSpan
       $Reading= $Reading & _CommGetString()
       $End = TimerDiff($Start)
       If $Reading <> '' Then $ReadStep = $ReadStep + 1
       If $ReadStep > 100 Then $End = $TimeSpan
       Sleep(0)
   WEnd
   Return $Reading
EndFunc
```

Code sample 17.2 _CommReadString, a new function for getting information from a serial port.

As for _CommSendStandard, add this function to the file CommMG.au3 so that it can be called by the subsequent scripts. Let us examine the "_CommReadString" function now. Its argument is $TimeSpan, which, as the name suggests, is a time span through which the function will work. For the examples below, this time span can be between less than 1 s and potentially more than 20 s. The value given for $TimeSpan needs to be higher than the maximum estimated time. Here, let us set it at 50 000 (= 50 s, as the unit is millisecond). Depending on the application, this value could be much higher. For example, there are analyzes in gas chromatography for which this time span can be longer than 1 h.

The following four lines in Code sample 17.2 present the variables used in the code. The first one is only declared, and no value is attributed to it. This is because functions in the CommMG.au3 library must be declared. The following ones do not interact directly with those functions, and thus are not declared, but simply assigned a value to each. The first of them, $Start, receives the value of the function "TimerInit," which can be assumed to be 0. The following two receive 0 each.

After variable allocation, a While loop starts, which continues while the $End value is smaller than $TimeSpan. In this loop, the variable $Reading receives its value (which is nothing) before the loop starts, plus the contents of _CommGetString. In the cases of the scripts being studied ahead, _CommGetString gets nothing most of the time. Therefore, each run of the loop $Reading is added of nothing (the addition is done using the concatenator, &, see Section 4.5 for another example). When _CommGetString receives some data, they are passed to $Reading. For example, suppose that at a run of the loop _CommGetString got "Messa." This word is passed to $Reading, which becomes "Messa." Suppose that in the next run _CommGetString got "ge." It will pass this to $Reading, which now will be "Message." After that, $End receives the value of the function "TimerDiff" in relation to $Start, which will be a fraction of second in the first run of the loop, but will increase steadily up to the value of $TimeSpan (50 s, in our case). All this is done at a sleep time of 0, which is not zero, but the minimum possible time that AutoIt can make. This very small step is necessary to ensure that no information is lost between runs of the loop.

Then comes an If … Then section in which the content of $Reading is tested. If it ceases being empty, the variable $ReadStep has its value increased by 1. After that, there is another If … Then statement testing if $ReadStep is larger than 100. If yes, $End receives the value of $TimeSpan. Finally, after the loop, the function returns the value of $Reading.

Let us see the purpose of the two sequential If … Then statements in detail. The idea is to enable a way by which _CommReadString finishes before 50 s. As explained at the beginning, 50 s is a maximum time, and if always used it may result in inefficiency and even make the script impractical if the real maximum time is much shorter, like 1 s only. Then, only if the conditions for the two If … Then in sequence are false, the script will take 50 s to be finished. If any useful information is recorded before that, the first If … Then will start the variable $ReadStep, which will have a maximum value here of 101, that is, the loop will run 101 times and after that $End will be assigned the loop-exit value of $TimeSpan. This number of steps, 101, is a maximum arbitrary value and was adapted to the situations of the two examples given below. It is possible that for other applications this value needs to be higher.

We now see an example of the control of devices connected to serial ports using AutoIt. Unfortunately, you need to have the hardware to see them working for you. Still, the principles outlined here work for any device utilizing the RS-232 communication protocol, as, for example, devices built using Arduino, a popular microcontroller that has been used by many hobbyists for making low-cost prototypes of electronic devices.

This example demonstrates how AutoIt can be used to control a valve actuator. This valve actuator is called *EHMA*, made by Vici. It is commonly used in gas chromatography setups. It can be switched between two positions, A and B. The manual for this device is available on http://http://www.vici.com/support/tn/tn413.pdf. There, you can find on page 4 the information about serial communication protocols. There is said that baud rate is 9600, number of bits is eight, there is one stop bit, and no flow control or handshaking. Also, on page 5, there is a list of commands that allow the control and monitoring of the valve controller. Here, we are going to use CP, which shows the current position of the valve controller (A or B), and GOn, which makes the actuator go to position n, where n can be A (GOA) or B (GOB). Let us see a demonstration of the control of this device using AutoIt (Code sample 17.3):

```
1) #include "commMG.au3"
2) Local $port = 5
3) Local $portError
4) Local $ValvePosition
5) _CommSendStandard($port,$portError,InputBox("Valve rotation","Type
GOA or GOB"))
6) Sleep(1000)
7) _CommSendStandard($port,$portError,"CP")
8) $ValvePosition = _CommReadString(50000)
9) MsgBox(0,"Valve position",$ValvePosition)
```

Code sample 17.3 Controlling the EHMA valve with AutoIt.

Remember to remove the line numbers, as they are not part of the code. The script in Code sample 17.3 will only do anything useful if you actually have the EHMA valve connected to port 5 on your computer. In all other cases, it will still open the Input box, and you will be able to type anything there. However, even if you type the correct values, that are GOA or GOB, it will eventually result in error, as it will not be able to connect with the device. Supposing you do have the EHMA connected to your computer on port 5, if you type GOA, the valve will turn to position A (or do nothing if it is already at position A) and after a short while the message "Position is A" will be shown. If you type GOB, the behavior is similar, but the valve should go to position B, and the message "Position is B" should be shown. The next paragraphs will examine the Code sample 17.3 in detail.

The first line of Code sample 17.3 consists in including the commMG.au3, because it contains the functions that we created, _CommSendStandard and _CommReadString, which are necessary to control and monitor the valve controller.

The next three lines in Code sample 17.3 declare three variables. $port is the port being accessed. It is shown as number 5, but on your computer this may be different. You must verify that otherwise the script will not work. Then, two more local variables are declared, $portError and $ValvePosition. As mentioned earlier, for the functions in commMG.au3 variables must be declared even if they do not receive an initial value. $portError is the type of error that the script will return in case something goes wrong. A full list is in the CommMG.au3 file.

After the variable declarations, the function _CommSendStandard (Code Sample 17.1) is called twice, with a 1s interval between the calls. The first call sends what we input as a command to the valve actuator. The next call sends the command "CP," which, according to the device's manual, will return the current position of the valve. In the next line, the variable $ValvePosition receives the contents generated by the function _CommReadString, which should be the response to the CP command in the previous line. The final line calls a message box to be displayed, showing the contents of the variable $ValvePosition.

17.3 Monitoring in Real Time

Some COM-port devices are designed for real-time monitoring, but lack an interface that allows such monitoring to be easily observed on the computer screen. Here, we will see a very simple solution for this problem using spreadsheet software. In order to implement the solution, it helps that you have studied Chapter 9, which presents the basic functions dealing with spreadsheets.

The idea here is to gather the data generated by an instrument connected to a COM port and pass it as data for a spreadsheet containing a plot showing the value of the data being collected in real time. Let us begin with an example which does not need any device connected to a COM port:

```
#include <OOoCalc.au3>

$path = "C:\YourFolder\"
$Book = _OOoCalcBookOpen($path&"Output.ods")

Sleep(2000)

While 1
    $sec = @SEC + 0
    _OOoCalcWriteCell($Book, $sec, 1,0)
    For $i = 30 to 2 step -1
        $up = _OOoCalcReadCell($Book, $i-1,0)+0
        _OOoCalcWriteCell($Book, $up, $i,0)
    Next
    Sleep(100)
WEnd
```

Code sample 17.4 Transferring data from the computer clock to a Calc spreadsheet.

Before running the script, make sure you create a Calc spreadsheet named "Output.ods" and save it in a folder ("C:\YourFolder\", in our example). The spreadsheet must have data on its first column (column index 0), the first row (index 0) containing a description of the data, and rows index 1 to 30 the numerical data, that is, the basis of the graph (Figure 17.1)

Close the spreadsheet file, and run the script. It will open the file and start adding values of the current second (time macro @SEC, see Chapter 11 for more on time macros) to the first data, at cell A2. Cells A3–A31 receive the value of the cells above them (the For loop in Code sample 17.4). The plot is made with cells A2–A31, and thus it is updated in real time. Code sample 17.4 should be easy to understand by now, as it has almost no new elements. It is arguably confusing, however, that we set the two numerical variables, "$sec" and "$up" as sums of a value plus 0. Of course, that such sum does not change the numerical value being passed, and thus looks unnecessary. However, I found that _OooCalcWriteCell will input values as strings, and not numbers, to Calc unless the value passed is explicitly a number. An easy way to assign a variable as a number in AutoIt is to add 0 to its assigned value, as done in Code sample 17.4.

The same strategy can be applied to display the data from an instrument connected to a COM port. Here, I demonstrate how this can be done for a flow meter (Restek ProFlow 6000), which is used to measure gas flows. This instrument can be connected to a computer via USB. Examination of its output to the computer (using a script containing the _CommReadString function defined in Code sample 17.2, for example) shows that the typical signal consists of a phrase like "Flow@40 mL/minCRLF" or "Flow@under rangeCRLF" or "Flow@over rangeCRLF." The value of 40 was only an example, the numbers

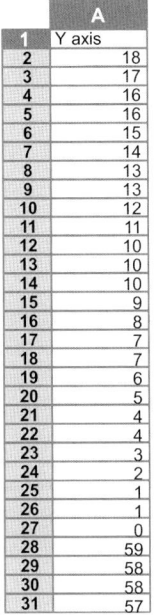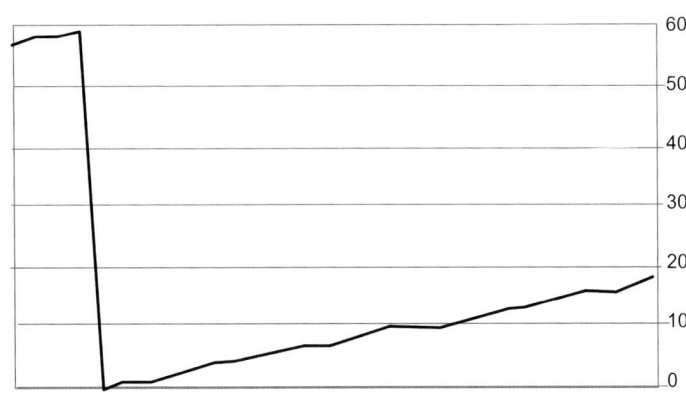

Figure 17.1 Calc spreadsheet with the column containing the data from the clock and the plot showing the data. The plot is updated every second or so as soon as new data are read using the script in Code sample 17.4.

varying between 0 and 500. CRLF here is not the group of characters themselves, but a symbol for line break. Therefore, the input must be separated at CRLF, @ and blank space (''), so that only the numeric values, or under or over are passed as inputs. Once we know how to deal with the data input, we can write a new function, which we can add to the commMG.au3 file:

```
Func ReadFlowMeter($period)
    $FlowReading = _CommReadString($period)
    $ReadingLine = StringSplit($FlowReading,@CRLF)
    $ReadingWords = StringSplit($ReadingLine[1]," ")
    $ReadingValue = StringSplit($ReadingWords[1],"@")
    If UBound($ReadingValue) <> 2 Then
        If $ReadingValue[2] = "under" Then
            $output = 0
        ElseIf $ReadingValue[2] = "over" Then
            $output = 500
        Else
            $output = $ReadingValue[2]
        EndIf
    Else
        $output = 0
    EndIf
    Return($output)
EndFunc
```

Code sample 17.5 Function to read the flow values from Restek Proflow 6000 flow meter.

This function processes the input generated by the flow meter by repeatedly splitting the string that comes as the standard input, and dealing with possible exceptions (under and over scale values). Now we can write a script that reads the flow from the flow meter and plots the readings on a Calc spreadsheet graph:

```
#include "commMG.au3"
#include <OOoCalc.au3>

Local $port = 4
Local $portError
_CommSetPort($port, $portError, 115200, 8, 0, 1, 2)
$path = "C:\YourFolder\"
$Book = _OOoCalcBookOpen($path&"FlowOutput.ods")
Sleep(1000)

While 1
    $flow = ReadFlowMeter(1000)+0
    _OOoCalcWriteCell($Book, $flow, 1,0)
    For $i = 30 to 2 step -1
        $up = _OOoCalcReadCell($Book, $i-1,0)+0
        _OOoCalcWriteCell($Book, $up, $i,0)
    Next
    Sleep(100)
WEnd
```

Code sample 17.6 Script that reads the flow from the Restek Proflow 6000 flow meter and plots the results on a Calc spreadsheet graph.

Note that the parameters for the COM port connection here are basically the same as those in the previous examples, except for the baud rate, which here is 115 200. Make sure you have the flow meter connected to your computer via a USB port and that the correct port number is in your code, and pay attention to file and folder names in the script. The rest of the code is virtually identical to the one in Code sample 17.5. If you have the device, you can run the script and should be able to read the flow of the flow meter on the plot (Figure 17.2).

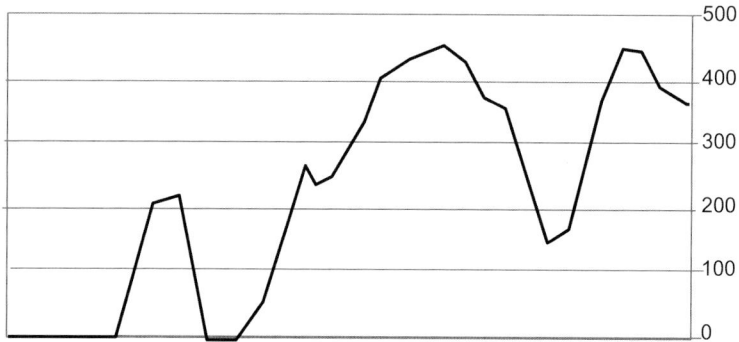

Figure 17.2 Calc spreadsheet graph showing the reading of the Restek Proflow 6000 flow meter obtained using the script in Code sample 17.6.

17.4 Implications for Other Devices

As you saw, to control devices directly using RS232 protocol is much more complex than simply using a software purposely designed for that. The codes in this chapter are just simple examples to demonstrate the control of devices using AutoIt. With the knowledge on the script strategies that you learned through the book, you should be able to include the commands _CommSendStandard and _CommGetString at the appropriate position in the code so that the device controlled interacts in a useful form with other devices. The approach outlined here should work for any other similar devices for which communication protocols are available from the manufacturer.

Finally, a few words about Arduino. Arduino is an open-source hardware platform that has become highly popular in the last 5 years or so. With Arduino, it became easier to build electronic devices from scratch. There are countless examples of devices built using Arduino on the Internet, and some of them could be adapted for use in the laboratory. In fact, there is a movement for open-source laboratories, which advocates that many instruments could be built by their end users. There are many occasions in which this type of approach can be useful. See more comments in Chapter 7.

Although it has become easier to build electronic devices since Arduino appeared, it does not mean that this is easy. Knowledge about electronics, and sometimes about mechanics, is necessary. It is much harder to learn this than to learn AutoIt with the pure intention of scripting. It is not the purpose of this book to teach Arduino, and thus I will only mention that, for the cases in which I used Arduino together with AutoIt (very simple custom-made devices), the communication protocols as in Code sample 17.1 (baud rate = 9600, bit number = 8, stop bits = 1, no parity, no flow control or handshake) worked. Interested readers can find plenty of information about Arduino on the Internet and in specialized books.

17.5 Other Technologies for Instrument Control

In addition to controlling instruments using COM ports, AutoIt has a library for GPIB/VISA, which mean, respectively, General Purpose Interface Bus and Virtual Instrument Software Architecture. The AutoIt library for them is contained in the file Visa.au3, which is an integral part of the AutoIt installation package. Although widespread among many instrument manufacturers, it is not common for the end user to need to deal with these technologies. Unless you are designing an instrument yourself, it is unlikely you will need to deal with them. Therefore, they are not covered in this book.

For readers interested in advanced topics related to instrument control, the reference at the end of the chapter can be a good starting point.

17.6 Summary

- Many instruments are controlled via USB ports connected to a computer.
- USB and serial ports employ the RS-232 protocol. These ports are also named COM ports.

- The most important parameters of the RS-232 protocol are: baud rate, number of bits, parity, stop bits, flow control, and handshaking.
- AutoIt has a library to deal with COM ports, but it needs to be downloaded separately.
- The most important functions in the COM port library are _CommSetPort, _CommSendString, and _CommGetString.
- In order to simplify their use, I created two new functions: _CommSendStandard and _CommReadString.
- The control of a valve actuator and the displaying of a flow meter reading were demonstrated as examples.
- Many other devices can be controlled using COM ports, including those built using Arduino, thus enabling the integration of devices built in the laboratory to standard devices.
- AutoIt also enables GPIB/VISA control, but this topic was not covered here.

Suggested Reading

Russo, M.F. and Echols, M.M. (1999) *Automating Science and Engineering Laboratories with Visual Basic*, John Wiley & Sons Inc., Brisbane, 355 pp.

18

Introduction to Graphical User Interface (GUI)

In addition to being able to control instruments with the computer, as learned in Chapter 17, it is essential that this control is made accessible to users without knowledge on AutoIt. This is where graphical user interfaces (GUIs) enter. AutoIt supports the design and creation of GUI, which makes it a complete solution for making software to control laboratory instruments.

Creating a GUI in AutoIt involves a little more coding than in the scripts presented so far, but it is not difficult. Perhaps the main obstacle when creating a GUI is the jargon. Although almost everybody is familiar with the elements of a GUI (buttons, text boxes that respond to mouse clicks, others that do not, etc.), only specialized programmers are usually familiar with the technical terms associated with these elements. This and the next chapter aim to make GUI creation with AutoIt as easy and simple as possible for the uninitiated reader by means of explaining the jargon and also by providing several different examples on how to utilize a few GUI elements. There are many other GUI elements that will not be covered; but, after the introduction provided here, you should be able to use the AutoIt help file and progress by yourself through the contents.

Before starting, it is important to mention that there are two ways of making a GUI in AutoIt: (i) by directly writing the code and (ii) by using a GUI creation utility, like Koda, which can be accessed using SciTE (either use Ctrl + m or go to Tools and select Koda form designer). Although using Koda seems to be the obvious path, I personally think that it is better to start by typing the code, so that you learn the important aspects of each GUI element. If we go straight to Koda, it can be difficult to follow what is being done. Therefore, Koda will be introduced only in Chapter 22, after the most important aspects of GUI creation have been covered in this and the following three chapters.

18.1 Making a Very Simple GUI

The essential function in a GUI code is GUICreate. As its name indicates, this function creates the GUI. Let us see a very simple example:

```
$Tinny = GUICreate("Tinny", 200, 100, 500, 200)
```

Code sample 18.1 Introducing GUICreate.

Run the script. You will see nothing happening. Still, the function GUICreate created a GUI that consisted in a window titled "Tinny" measuring 200 pixels in width, 100 pixels in height, and positioned at the coordinates 500 × 200 on the screen. In order to see the window, however, you must call the function GUISetState with no arguments, as shown in the code below:

```
$Tinny = GUICreate("Tinny", 200, 100, 500, 200)
GUISetState()
```

Code sample 18.2 Introducing GUISetState.

Run the script. You did not see the window, but the function GUISetState worked. The problem is that the window appeared only for a moment, and thus you did not see it. In order to solve this issue, we can set a sleep time, as in the code below:

```
$Tinny = GUICreate("Tinny", 200, 100, 500, 200)
GUISetState()
Sleep(3000)
```

Code sample 18.3 Making a window visible for 3 s.

Now run the script. Finally, you can see the window (Figure 18.1)! You can move it and even minimize it, but you need to be quick, because after 3 s it is gone; however, you cannot close it by clicking on the closing button. A window like that is not very useful. Windows controlling instruments should stay visible indefinitely. In order to do that, we set an infinite loop containing a condition that allows us to close that window. This is done as in the following code:

```
#include <GUIConstantsEx.au3>
$Tinny = GUICreate("Tinny", 200, 100, 500, 200)
GUISetState()
While 1
    If GUIGetMsg() = $GUI_EVENT_CLOSE then Exit
WEnd
```

Code sample 18.4 Making a window that lasts as long as we wish.

Run the script. The window now should last for as long as we wish, as we are able to close it by clicking the close button. Code sample 18.4 has several different elements compared to Code sample 18.3. First, now the file "GuiConstantsEx.au3" is included. As the name implies, this file contains several constants that are useful when working with GUIs. One of these constants is $GUI_EVENT_CLOSE. This constant is the event

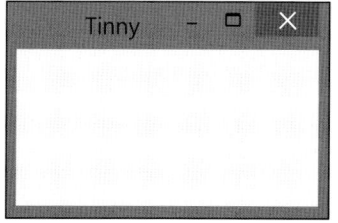

Figure 18.1 GUI generated with Code sample 18.3.

of closing the window, which is triggered by clicking on the close button. In the code, this constant is used like this. There is the While loop, which here is an infinite loop. It says "While 1". Because 1 will always be 1, the loop will never end. Therefore, the actions inside the loop should repeat forever. In this case, the action is the test if the value returned by the function GUIGetMSG is equal to $GUI_EVENT_CLOSE. If it is, then the command Exit is called, and the script is finished, and the GUI disappears. GUIGetMSG is a very important function that is watching what is being done with the GUI. It is using this function that we can make GUIs do useful work.

18.2 Adding Simple Elements to a GUI

The GUI that we created in the previous section has no use, since the only element in it is the close button. Let us add a few elements to it so that it becomes more useful. Suppose we have a simple instrument, like an automated dilutor, and that we know how to use communication protocols to control it using a computer. In this case, the first thing to do is to conceptualize the graphical part of the GUI. A drawing is very helpful (Figure 18.2).

The elements of this GUI are: a title (Fake Dilutor), two labels (Original … and Dilution …), two input areas beside each label, and a button. Below goes a preliminary code for such a GUI:

```
1) #include <GUIConstantsEx.au3>
2) $FADI = GUICreate("FakeDilutor", 200, 150, 500, 200)
3) $Label1 = GUICtrlCreateLabel("Original solution volume (between
0.1 and 1):",20,20,140,25)
4) $Input1 = GUICtrlCreateInput("",150,20,25,20)
5) $Label2 = GUICtrlCreateLabel("Dilution water volume (between 0.1
and 10):",20,60,140,25)
6) $Input2 = GUICtrlCreateInput("",150,60,25,20)
7) $Button1 = GUICtrlCreateButton("Go", 50, 100, 100, 25)
8) GUISetState()
9) While 1
10)     If GUIGetMsg() = $GUI_EVENT_CLOSE then Exit
11) WEnd
```

Code sample 18.5 A GUI with the graphical elements displayed in Figure 18.1.

Remove the line numbers, run this script, and you should see a GUI that resembles the diagram in Figure 18.2 (Figure 18.3).

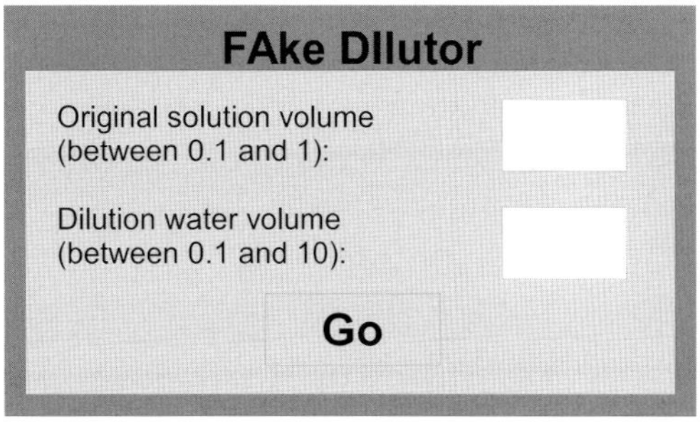

Figure 18.2 Sketch of a GUI for an automated dilutor.

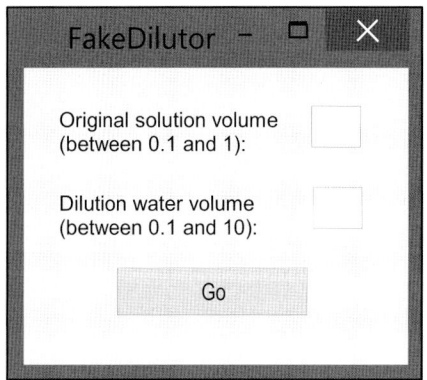

Figure 18.3 GUI created with Code sample 18.5.

As with the GUI in Code sample 18.4, you can move it, click on it, and close it. In addition, you can type in the input boxes. However, nothing really happens when you click the "Go" button. Therefore, this GUI only contains the graphical elements and no code attached to them. Still, let us examine the Code sample 18.5. The difference between it and Code sample 18.4 is that now several new variables were declared. These variables are the graphical elements of the GUI. Their positions and sizes are set in the arguments for the functions creating them (GUICtrlCreateLabel, GUICtrlCreateInput, and GUICtrlCreateButton). These functions work in a very similar way to the GUICreate function, which was already explained: the first argument is a text, which is displayed on each element; the following four arguments are numbers for position and size of each element, all of them in pixels. Note that the reference position of each element is the top left corner in the GUI main form. Note that for GUICtrlCreateInput the text field was left blank (""), as it is probably more appropriate in this case. However, it is possible to add any text as an argument in the case this has a useful purpose.

Let us now add a functionality to this GUI. We are going to enable the calculation of the dilution achieved by entering values to the inputs and displaying the result when we click on the Go button. To do that, we need to add some lines of code to our GUI. First, we create a function that does the dilution calculation:

```
1)  Func CalculateDilution($vol1,$vol2)
2)      If $vol1 >= 0.1 and $vol1 <= 1 and $vol2 >= 0.1 and $vol2 <= 10
3)      Then
4)          $total = $vol1 + $vol2
5)          $dilution = $total / $vol1
6)          MsgBox(0,"Now diluting","Final volume: "&$total&@CR&"Dilution: "&$dilution)
7)      Else
8)          MsgBox(16,"Error","Please input correct values")
9)      EndIf
10) EndFunc
```

Code sample 18.6 A function that calculates the dilution in a mixture of two liquids.

Save this code without the line numbers as a file named FADIfunctions.au3. Now write the following code:

```
1) #include <GUIConstantsEx.au3>
2) #include <FADIfunctions.au3>
3) $FADI = GUICreate("FakeDilutor", 200, 150, 500, 200)
4) $Label1 = GUICtrlCreateLabel("Original solution volume (between
0.1 and 1):",20,20,140,25)
5) $Input1 = GUICtrlCreateInput("",150,20,25,20)
6) $Label2 = GUICtrlCreateLabel("Dilution water volume (between 0.1
and 10):",20,60,140,25)
7) $Input2 = GUICtrlCreateInput("",150,60,25,20)
8) $Button1 = GUICtrlCreateButton("Go", 50, 100, 50, 25)
9) GUISetState()
10) While 1
11)     Switch GUIGetMsg()
12)         Case $GUI_EVENT_CLOSE
13)             Exit
14)         Case $Button1
15)             CalculateDilution(GUICtrlRead($Input1),
GUICtrlRead($Input2))
16)     EndSwitch
17) WEnd
```

Code sample 18.7 GUI with a functional button.

Run this script (remembering of removing the line numbers), and now you should be able to type in the volumes for dilution and, when pressing the Go button, to see the result of the calculation. You may have noted that Code sample 18.7 is based on Code sample 18.5. A difference between the two versions is the inclusion of the FADIfunctions.au3 file in Code sample 18.7, which makes it possible to access the function created in Code sample 18.6. Another difference is the content of the While loop in the two versions. In Code sample 18.5, there was only a brief If … Then statement inside the loop. Here, there is a Switch statement instead. As it was explained in Section 14.2, it is composed of the word Switch, followed by a variable, in this case the return of the function GUIGetMsg. Then, there is a line with the word Case, followed by the first possible value of the variable. In this case, this value is $GUI_EVENT_CLOSE, below which is the action to be done (the Exit function). The next Case is followed by $Button1, which is the other possible input for this GUI. In this case, on pressing the Go button, the actions listed below this line until the next Case or the word EndSwitch are performed. In this case, the action is the CalculateDilution function using the inputs of the GUI. The Switch structure is similar to the If … Then structure, but more convenient if there are many options. We could have easily used an If … Then in Code sample 18.7 instead, but usually a GUI has more than one button, which would make the code very complex using If … Then. Finally, you can now compile your script so that it can work as a stand-alone program. On SciTE, instead of pressing F5 and running the script, press Ctrl + F7, and save your file as .exe. Now you can run your new .exe file as an independent software.

18.3 Setting Keyboard Shortcuts

Some users prefer keyboard shortcuts to mouse clicks when controlling their software. Also, as we know, if you are going to automate different programs with AutoIt, it is better to rely on keyboard shortcuts than on mouse clicks. In AutoIt, keyboard shortcuts

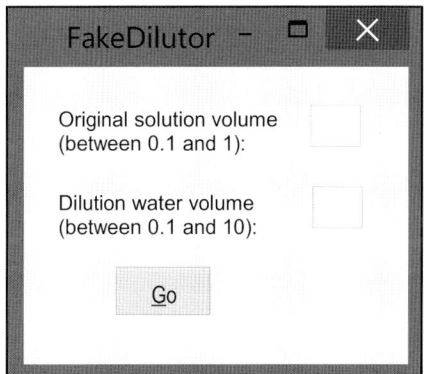

Figure 18.4 GUI created with Code sample 18.8.

can be of two kinds: hotkeys and accelerators. For a GUI, the most useful are normally accelerators, because they only work for the specific GUI. For example, if you set an accelerator that is activated using Ctrl + a, it will do an action only inside the GUI; in other words, if you activate another application and send Ctrl + a, this shortcut will do what is expected to do inside this application. However, if you set a hotkey instead, pressing Ctrl + a on another application will still activate the action on the GUI that you created. Therefore, hotkey are set to be used as general shortcuts that have the purpose of performing their task, no matter which application is running in the foreground. Instead, accelerators are meant to a specific application. Let us now include an accelerator in our GUI (Code sample 18.8):

```
1) #include <GUIConstantsEx.au3>
2) #include <FADIfunctions.au3>
3) $FADI = GUICreate("FakeDilutor", 200, 150, 500, 200)
4) $Label1 = GUICtrlCreateLabel("Original solution volume (between 0.1 and 1):",20,20,140,25)
5) $Input1 = GUICtrlCreateInput("",150,20,25,20)
6) $Label2 = GUICtrlCreateLabel("Dilution water volume (between 0.1 and 10):",20,60,140,25)
7) $Input2 = GUICtrlCreateInput("",150,60,25,20)
8) $Button1 = GUICtrlCreateButton("&Go", 50, 100, 50, 25)
9) Local $Accel[1][2] = [["^g", $Button1]]
10) GUISetAccelerators($Accel)
11) GUISetState()
12) While 1
13)     Switch GUIGetMsg()
14)         Case $GUI_EVENT_CLOSE
15)             Exit
16)         Case $Button1
17)             CalculateDilution(GUICtrlRead($Input1), GUICtrlRead($Input2))
18)     EndSwitch
19) WEnd
```

Code sample 18.8 GUI with a functional button that responds to Ctrl + g.

Remove the line numbers and run the script. Note that now the "G" on Go is underlined, which was achieved on the code by defining the button label as "&Go". The & symbol, in this case, determines that the next character comes underlined (Figure 18.4).

Type in the values in the input fields and, instead of pressing the Go button, send Ctrl + g. It will do the calculation as if you have clicked. It should work as in the previous code. The function that set the accelerator Ctrl + G was "GUISetAccelerators," which used as argument the variable "$Accel." This variable was declared on the previous line as an array having one line and two columns. The elements of the array are "^g" and $Button1. The first element, ^g, means Ctrl + g. If you preferred to use Alt + g, it would be "!g" instead. Finally, note that before $Accel it was necessary to write "Local". This is necessary when dealing with arrays, see Section 8.1.

18.4 Summary

- AutoIt enables the creation of GUIs.
- GUIs are very useful to enable the interaction of users with software.
- There are two ways of creating GUIs in AutoIt: by directly typing the code and by using Koda, a graphical editor.
- GUICreate, GUISetState, and GUIGetMsg must be all used together in order to create a proper GUI.
- $GUI_EVENT_CLOSE is a global variable that enables the use of a closing button in a GUI.
- Before creating a GUI, it is useful to make a sketch of it.
- GUICtrlCreateLabel creates a label.
- GUICtrlCreateInput creates a text input area.
- GUICtrlCreateButton creates a button. Usually, you associate a click on a button to an action, which is defined in a function.
- The GUI displays its elements, and some of them can be associated with actions. The function GUIGetMsg is the key function that pools the interface constantly. If a certain action is taken, this function recognizes it.
- A Switch … Case conditional is often used to allow GUIGetMsg to choose among several different possible actions.
- It is possible to set up keyboard shortcuts for your GUI by using accelerators.

19

Using GUI to Control Instruments

You have learned in Chapter 18 the very basic aspects of graphical user interface GUI creation using AutoIt. Here, we will extend that knowledge by making GUIs that control real instruments. Also, in Chapter 17, you learned how to directly control instruments using AutoIt via COM ports. In this chapter, we will make GUIs that enable the easy control of such instruments for the end user, without needing to access the scripting code. As in Chapter 17, you should ideally have access to the instruments being controlled to see the scripts working. If you do not, however, the examples shown here can still be useful if you plan to control similar devices.

19.1 GUIs to Control the EHMA Valve Actuator

In Chapter 17, a simple script (Code sample 17.3) was introduced to control the EHMA valve actuator. In a sense, it was a GUI: A pop-up window asked the user to type a command (GOA or GOB) in order to turn the actuator to one of the two possible positions, A or B. Note that it was possible for the user to type anything else. While this would not be a problem for this specific case (anything else typed would simply result in the actuator doing nothing), in other situations it is possible that by sending the wrong command an undesirable action will result. Therefore, let us see some options of GUI that would limit the actions of the user.

A familiar way of allowing only a few options to the user is by using "radio buttons." Let us use radio buttons in a GUI to control the EHMA valve actuator:

```
#include <GUIConstantsEx.au3>
#include "commMG.au3"
Local $port = 5
Local $portError
$CrioControl = GUICreate("EHMA", 200,100,500,500)
$radio1 = GUICtrlCreateRadio("Position A",20,20)
$radio2 = GUICtrlCreateRadio("Position B",20,40)
$button1 = GUICtrlCreateButton("Go",20,80,40,20)
GUISetState()
```

Practical Laboratory Automation: Made Easy with AutoIt, First Edition. Matheus C. Carvalho.
© 2017 Wiley-VCH Verlag GmbH & Co. KGaA. Published 2017 by Wiley-VCH Verlag GmbH & Co. KGaA.

```
While 1
   Switch GUIGetMsg()
       Case $GUI_EVENT_CLOSE
           Exit
       Case $button1
           If GUICtrlRead($radio1) = 1 Then
               _CommSendStandard($port,$portError,"GOA")
           ElseIf GUICtrlRead($radio2) = 1 Then
               _CommSendStandard($port,$portError,"GOB")
           EndIf
   EndSwitch
WEnd
```

Code sample 19.1 GUI with radio buttons to control the EHMA valve actuator.

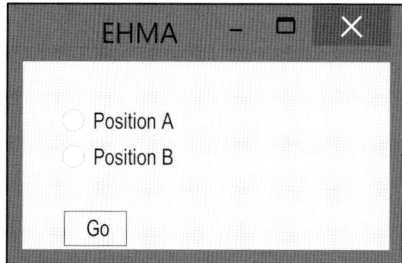

Figure 19.1 GUI with radio buttons to control the EHMA valve actuator.

If you run the script, you should see a window like this (Figure 19.1). The radio buttons are the two white circles near the options "Position A" and "Position B." If you do not have the EHMA actuator connected to your computer using COM port 5, the script will not work properly, as we learned in Chapter 17. If you do have it, you will see that the valve will move to position A if you choose Position A, and to position B if you choose Position B.

Now let us see the Code sample 19.1 in detail. Overall, it is quite similar to the codes presented in Chapter 18, with the addition of the necessary parts to include COM port control (commmg.au3). Inside the While loop, as in Chapter 18, there are some actions that are triggered if the "Go" button ($button1) is pressed. Here comes the main difference in relation to the previous scripts, that is, the use of the radio buttons. The function GUICtrlRead is called for $radio1, which is the radio button correspondent to Position A. If this radio button had been selected (by clicking on it, turning it black), pressing the Go button the actuator should turn the valve to position A. If the other radio button had been selected, the actuator would turn to position B. AutoIt determines that a radio button was clicked by assigning it a value of 1, which is the reason why in the code we test for GuiCtrlRead($radio) = 1.

Now let us see a variation of the same theme. Another way to give restricted options to a user is by means of combo boxes, which are expanding lists containing options. See the code below:

19.2 Controlling Two or More COM Ports in the Same Script

```
#include <GUIConstantsEx.au3>
#include "commMG.au3"
Local $port = 5
Local $PortError
$CrioComboControl = GUICreate("EHMA", 200,100,500,500)
$button1 = GUICtrlCreateButton("Go",20,80,60,20)
$combo1 = GUICtrlCreateCombo("",20,20,60,20)
GUICtrlSetData($combo1, "PosA|PosB", "PosA")
GUISetState()
While 1
   Switch GUIGetMsg()
      Case $GUI_EVENT_CLOSE
         Exit
      Case $button1
         Switch GUICtrlRead($combo1)
            Case "PosA"
               _CommSendStandard($port, $PortError, "GOA")
            Case "PosB"
               _CommSendStandard($port, $PortError, "GOB")
         EndSwitch
   EndSwitch
WEnd
```

Code sample 19.2 GUI with a combo box to control the EHMA valve actuator.

Figure 19.2 GUI with a combo box to control the EHMA valve actuator.

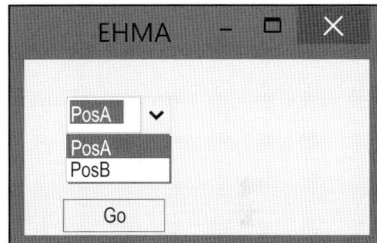

The GUI for the script with the combo box should look like (Figure 19.2):

Running the script, you should see the window as in Figure 19.2. If you have EHMA connected to your COM port 5, by selecting PosA and pressing the Go button, it should turn the valve to position A, while by selecting PosB and pressing the Go button it should turn the valve to position B. The code (Code sample 19.2) is very similar to that of the radio buttons (Code sample 19.1), the main difference being that instead of creating two controls, only one was created, and thus only the state of this control is tested.

19.2 Controlling Two or More COM Ports in the Same Script

By now, you know that you can synchronize several instruments each one with its GUI by writing scripts as those presented in Chapters 3–5 or 11–16. However, sometimes an analytical setup consists of more than a single device connected to the computer, and thus if all such devices could be controlled by a single GUI, this would make things simpler for the users, especially if the devices being controlled are not supported by GUIs, as it is the case of the EHMA. Suppose, for example, that we have two EHMA working in concert. Let us see a simple GUI to control such a system:

```
#include <GUIConstantsEx.au3>
#include "commMG.au3"
Local $valco1port = 4
Local $valco2port = 5
Local $PortError
$CrioValcoControl = GUICreate("EHMA", 200,100,500,500)
$button1 = GUICtrlCreateButton("Valco1",20,80,60,20)
$button2 = GUICtrlCreateButton("Valco2",100,80,60,20)
GUIStartGroup()
$radio1 = GUICtrlCreateRadio("Position A",20,20)
$radio2 = GUICtrlCreateRadio("Position B",20,40)
GUIStartGroup()
$radio3 = GUICtrlCreateRadio("Position A",100,20)
$radio4 = GUICtrlCreateRadio("Position B",100,40)
GUISetState()
While 1
   Switch GUIGetMsg()
      Case $GUI_EVENT_CLOSE
         Exit
      Case $button1
         If GUICtrlRead($radio1) = 1 Then
            _CommSendStandard($valco1port, $PortError, 1)
         ElseIf GUICtrlRead($radio2) = 1 Then
            _CommSendStandard($valco1port, $PortError, 2)
         EndIf
      Case $button2
         If GUICtrlRead($radio3) = 1 Then
            _CommSendStandard($valco2port, $PortError, "GOA")
         ElseIf GUICtrlRead($radio4) = 1 Then
            _CommSendStandard($valco2port, $PortError, "GOB")
         EndIf
   EndSwitch
WEnd
```

Code sample 19.3 GUI that controls two EHMA valve actuators.

The GUI in Code sample 19.3 should look like (Figure 19.3):

This script is very similar to the one in Code sample 19.1, the main differences are that now we need to allocate two variables to two different COM ports, and that there are two groups each containing two radio buttons. By expanding Code sample 19.3, it is possible to control several devices each connected to a different COM port (Figure 19.3).

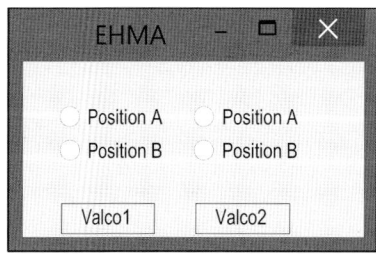

Figure 19.3 GUI that controls two EHMA valve actuators.

19.3 A GUI to Control a Digital Balance

It is fair to say that the GUIs that were presented so far in this chapter are exceedingly simple and not real-life examples of what is used in laboratories. Therefore, let us now create a more elaborate GUI. The purpose of this GUI is to control a digital balance. The manufacturer of the chosen balance (Mettler-Toledo) made available the set of instructions necessary to enable communication between the balance and a computer. Two manuals are useful for this purpose. The first can be downloaded from http://http://us.mt.com/dam/mt_ext_files/Editorial/Generic/2/XS_Precision_BA_Editorial-Generic_1097233026453_files/xs-prec-ba-e-11780659c.pdf.

In this manual, go to page 40 and find the settings for serial communication. They are: 9600 of baud rate, eight data bits, no parity, one stop bit, and Xon/Xoff flow control. Also, it shows that the end-of-line character is <CR><LF>, and the ANSI/WINDOWS character set is used. These configurations can be modified (procedure explained in the manual). In our example, let us modify the flow control to none, and the end of line character to CR. This way, our _CommSendStandard function can be used directly. So, if you have any of the balances that are in the family associated with the manual, you should do these modifications in order that the scripts presented here work. The ANSI/WINDOWS character set is the default and does not need to be modified.

The other manual that you need to access is on http://http://us.mt.com/dam/mt_ext_files/Editorial/Generic/5/AX_MX_UMX_SICS_0x0002467000027246000a4b7e_files/ax-mx-umx-sics-ba-11780417.pdf.

This is a complete reference for the commands that can be sent via RS232 protocol to the balances. Here, we will use only three of the listed commands: (i) "SI," which gives an instantaneous reading of the balance; (ii) "S," which gives a stable reading of the balance; and (iii) "T," which tares the balance. Also, our GUI will enable the storage of the weighed data in a way that can be easily accessed later by a spreadsheet software like Microsoft Excel or LibreOffice Calc, for example. Let us see the code for such a GUI:

```
#include "commMG.au3"
#include <FileConstants.au3>
#include <GUIConstantsEx.au3>
#include <ColorConstants.au3>
#include <StaticConstants.au3>
Local $BalancePort
Local $PortError
$StoringFile = "C:\balance\BalanceReading.txt"
$ToledoBalance = GUICreate("Balance controller", 400, 100, 500, 200)
$TareButton = GUICtrlCreateButton("Tare", 20, 60, 50, 20)
$WeighButton = GUICtrlCreateButton("Weigh", 20, 80, 50, 20)
$ReadButton = GUICtrlCreateButton("Read balance", 160, 0, 80, 20)
GUICtrlCreateLabel("Balance reading:",80,20,80,20)
$Reading = GUICtrlCreateLabel("",160,20,80,18,$SS_RIGHT)
GUICtrlSetBkColor($reading,$COLOR_WHITE)
GUICtrlCreateLabel("Sample name:",80,40,80,20)
$SampleName = GUICtrlCreateInput("",160,40,80,20)
$StoreButton = GUICtrlCreateButton("Store", 160, 60, 80, 20)
$ChangeButton = GUICtrlCreateButton("Change file", 160, 80, 80, 20)
$COMButton = GUICtrlCreateButton("Set COM port", 300, 0, 80, 20)
GUICtrlCreateLabel("COM port:",250,20,80,20)
$COMport = GUICtrlCreateInput("",300,20,80,20)
GUISetState()
```

```
While 1
    Switch GUIGetMsg()
        Case $GUI_EVENT_CLOSE
            Exit
        Case $COMButton
            $BalancePort = GUICtrlRead($COMport)
        Case $ReadButton
            _CommSendStandard($BalancePort,$PortError,"SI")
            $BalanceReading = _CommReadString(1000)
            $ReadingArray = StringSplit($BalanceReading," ")
            GUICtrlSetData($Reading,$ReadingArray[4])
        Case $TareButton
            _CommSendStandard($BalancePort,$PortError,"T")
        Case $WeighButton
            _CommSendStandard($BalancePort,$PortError,"S")
            $BalanceReading = _CommReadString(50000)
            $ReadingArray = StringSplit($BalanceReading," ")
        Case $ChangeButton
    $StoringFile = FileSaveDialog("Type new file name","C:\","(*.txt)")
        Case $StoreButton
            $Name = GUICtrlRead($SampleName)
          FileWrite($StoringFile,$Name &" "& $ReadingArray[4] & @CRLF)
    EndSwitch
WEnd
```

Code sample 19.4 GUI that controls a digital balance.

The GUI controlling the balance looks like (Figure 19.4):

Running the script, you should see the GUI as depicted in Figure 19.4. Before analyzing the code, let us play with the GUI and see what it does. Again, it will only work completely if you have the balance connected to the correct COM port. This time, there is a button which enables the user can set the COM port number by typing it and pressing the "Set COM port" button. This must be the first action taken by the user to make the script work properly. Also, there is a "Change file" button that allows the user to choose the file, where he/she will store the reading data. It needs to be a text (.txt) file. Files of this type can be read by spreadsheet software like Excel or Calc. The other buttons on the GUI only do something correct if the balance is connected to the computer. If not, only error messages will be displayed. Supposing the balance is connected, "Tare" will tare the balance, "Weigh" will do the stable weight reading (showing the value in the "Balance reading" field), and "Read balance" will show the instantaneous reading of the balance in the Balance reading field. A sequence of actions to properly weigh a sample would be: (i) tare the balance by pressing Tare; (ii) put the sample on the balance; (iii) press Weigh to get the weight of the sample; (iv) type the sample name on the "Sample name" field; (v) store the data in the storing file by pressing "Store."

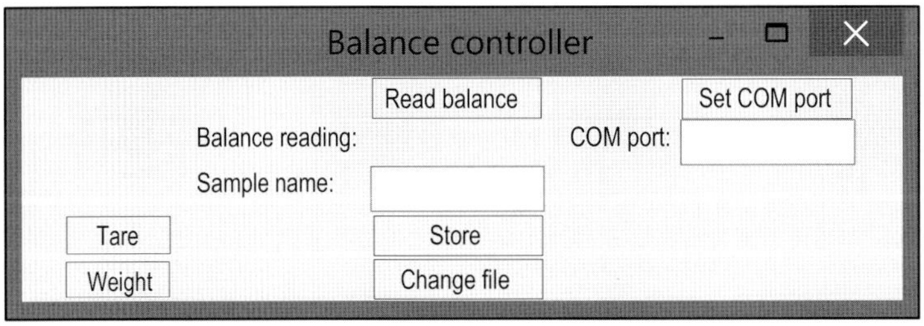

Figure 19.4 GUI that controls a digital balance.

Let us examine Code sample 19.4. By now, if you have been following the book chapter by chapter, it should be relatively easy to understand what is going on in the code: some libraries were included in order to enable some functions to work; variables related to COM ports were declared as this is required by commMG.au3; a variable, $StoringFile, receives a value containing the path of the standard file where the weighing data are saved; GUI elements were defined; the While loop is called, containing the actions that happen for each button that is clicked. Most actions are straightforward, but let us examine some that could be confusing. Two functions related to file manipulation were used: "FileSaveDialog" and "FileWrite." As their names suggest, one opens the file-saving window, that allows the user to choose the folder and file name of the file storing the weighing data. By default, C:\, the root folder, is suggested, and also the file type (.txt). FileWrite works by sending the data to the chosen file. The @CRLF macro ensures that a new line is started after the data are sent, so that each reading goes to a single line. Also, if the script is started again a different time, for example, after many days, the data will be stored at the end of the file as default, so nothing is overwritten. You should be familiar with StringSplit, but if not check Section 8.2.

Although functional, the GUI presented in Code sample 19.4 arguably looks messy and difficult to understand. Let us improve the presentation and usability of this GUI by using tabs:

```
1)  #include "commMG.au3"
2)  #include <FileConstants.au3>
3)  #include <GUIConstantsEx.au3>
4)  #include <ColorConstants.au3>
5)  #include <StaticConstants.au3>
6)  Local $BalancePort
7)  Local $PortError
8)  $StoringFile = "C:\balance\BalanceReading.txt"
9)  $ToledoBalance = GUICreate("Balance controller",200,160,500,200)
10) GUICtrlCreateTab(10, 10, 180, 140)
11) GUICtrlCreateTabItem("COM port")
12) $COMButton = GUICtrlCreateButton("Set COM port", 20, 40, 80, 20)
13) GUICtrlCreateTabItem("Weigh")
14) $ReadButton = GUICtrlCreateButton("Read", 20, 40, 50, 20)
15) $Reading = GUICtrlCreateLabel("",80,40,80,18,$SS_RIGHT)
16) GUICtrlSetBkColor($reading,$COLOR_MEDGRAY)
17) $TareButton = GUICtrlCreateButton("Tare", 20, 60, 50, 20)
18) $WeighButton = GUICtrlCreateButton("Weigh", 20, 80, 50, 20)
19) $StoreButton = GUICtrlCreateButton("Store", 20, 100, 50, 20)
20) GUICtrlCreateTabItem("Options")
21) $ChangeButton =GUICtrlCreateButton("Change file", 20, 40, 80, 20)
22) GUISetState()
23) While 1
24)     Switch GUIGetMsg()
25)         Case $GUI_EVENT_CLOSE
26)             Exit
27)         Case $COMButton
28)             $BalancePort = InputBox("Set COM port","Type in COM port number, check Control Panel, Hardware settings.")
29)         Case $ReadButton
30)             _CommSendStandard($BalancePort,$PortError,"SI")
31)             $BalanceReading = _CommReadString(1000)
32)             $ReadingArray = StringSplit($BalanceReading," ")
33)             GUICtrlSetData($Reading,$ReadingArray[4])
34)         Case $TareButton
35)             _CommSendStandard($BalancePort,$PortError,"T")
```

```
36)            Case $WeighButton
37)                _CommSendStandard($BalancePort,$PortError,"S")
38)                $BalanceReading = _CommReadString(50000)
39)                $ReadingArray = StringSplit($BalanceReading," ")
40)                GUICtrlSetData($Reading,$ReadingArray[4])
41)            Case $ChangeButton
42)                $StoringFile = FileSaveDialog("Type new file name","C:\","(*.txt)")
43)            Case $StoreButton
44)                $Name = InputBox("Store your data","The weight of this sample is " & $ReadingArray[4] & ". Add a sample name below:")
45)        FileWrite($StoringFile,$Name &" "& $ReadingArray[4] & @CRLF)
46)        EndSwitch
47) WEnd
```

Code sample 19.5 Improved GUI for controlling a digital balance.

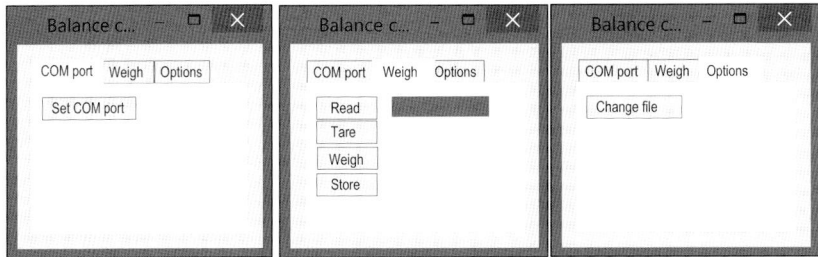

Figure 19.5 Improved GUI for controlling a digital balance.

The GUI controlling the balance looks like Figure 19.5, depending on the selected tab. (COM port, Weigh, and Options), Figure 19.5. I find this arrangement much easier to follow than the previous one (Figure 19.4). Relative to the coding, the main difference is that now the different groups of items are listed near their respective tabs (each tab marked by the "GUICtrlCreateTabItem" function). Note that it was necessary to call the "GUICtrlCreateTab" before that in the code. Also, in this code, a different set of constants, color constants, was included in order to enable the background of the "$Reading" field to be different from the background of the tab (function "GUICtrlSetBKColor"). A list of color constants can be found using the AutoIt help file.

19.4 Summary

- By combining GUI creation code with COM port control code, it is possible to create user interfaces for instruments.
- There are several options when creating a GUI. In the examples provided (Code samples 19.1 and 19.2), both radio buttons (GuiCtrlCreateRadio) and combo boxes (GUICtrlCreateCombo) could be used for the same purpose.

- It is possible to control more than one COM port in a single GUI. When doing so, it is useful to divide the GUI elements into groups by using the function GUIStartGroup.
- When reading data from an instrument, it is often necessary to cope with raw data that are not ready for use. In this case, string functions are very useful.
- A way to improve the presentation of a GUI is by dividing it into tabs, which is done using the functions GuiCtrlCreateTab and GuiCtrlCreateTabItem.

20

Multitasking GUIs

The approach to graphical user interface (GUI) creation presented in Chapters 18 and 19 is called the *message loop* approach in technical terms. This type of GUI works well for simple applications, but it has the important limitation that only a single action can be performed at a given time. This limitation can be a problem in some scenarios as, for example, when constantly reading the input from a measuring instrument. In this chapter, the other approach for GUI creation in AutoIt, the "event mode" approach, is presented. By using this approach, the script can perform more than one action at the same time.

20.1 The "GUIOnEventMode" Option

When writing the code for a GUI in AutoIt, it is possible to call the option "OnEventMode." If this is done, the whole code structure needs to be changed, as it is going to be seen. Let us start with a version of our very simple GUI of Code sample 18.4:

```
#include <GUIConstantsEx.au3>
Opt("GUIOnEventMode",1)
GUICreate("TinnyOnEvent", 200, 100, 500, 200)
GUISetOnEvent($GUI_EVENT_CLOSE, "CloseButton")
GUISetState()
while(1)
    sleep(500)
WEnd
Func CloseButton()
    Exit
EndFunc
```

Code sample 20.1 Version of Code sample 18.4 using the "OnEventMode" option.

If you run this script, you will see that the GUI is 100% identical to the one that is created using Code sample 18.4. However, it is clear that the codes are very different. First, the length: Code sample 18.4 was six-line long; this one is 11. Second, here there is the line Opt ("GUIOnEventMode," 1), which was absent in Code sample 18.4. In fact, you could rewrite Code sample 18.4 and add a line with Opt ("GUIOnEventMode," 0), and the script would perform exactly in the same way. The "GUIOnEventMode" option, if called with the parameter equal to 0, means that we are using the message loop approach for GUI creation. If called with the parameter equal to 1, then it means that we are using the event mode approach for GUI creation. The following line (GUICreate …) is

Practical Laboratory Automation: Made Easy with AutoIt, First Edition. Matheus C. Carvalho.
© 2017 Wiley-VCH Verlag GmbH & Co. KGaA. Published 2017 by Wiley-VCH Verlag GmbH & Co. KGaA.

identical to that in Code sample 18.4, but the next line, which uses the function "GUISetOnEvent," is absent in Code Sample 18.4. GUISetOnEvent is a fundamental piece of code when using event mode, and will be examined in detail.

In Code sample 20.1, the parameters for the GUISetOnEvent are "$GUI_EVENT_CLOSE," which is a constant that refers to the close button on the GUI, and "CloseButton," which is a function, that is, defined at the end of our code (the final three lines; it simply calls "Exit," which is a function that exits the GUI). Therefore, the first parameter for GUISetOnEvent is a GUI element, and the second a function. Note, however, that the function is called within quotes (" "), meaning that it is taken as a string. For practical purposes, this means that the function called by GUISetOnEvent needs to be written in the same file of the rest of the code. In other words, you cannot use an #include statement to call functions defined in different files. Therefore, in our example, you could not simply use the function Exit in the call for GUISetOnEvent; you needed to wrap it with another function defined in your code. This approach brings the limitation that functions that use parameters themselves cannot be directly called using GUISetOnEvent.

The remaining elements of Code sample 20.1 are familiar by now. The function GUISetState is essential when making a GUI, and therefore is present. Finally, there was a While loop. This While loop, which in the codes in Chapters 18 and 19 was the heart of the code, here merely keeps the window open by means of a sleep.

Therefore, the basic structure for a GUI code when using event mode is: including necessary files, declaring variables, defining GUI elements with their respective GUISetOnEvent functions, calling GUISetState, calling an infinite While loop, and finally defining the functions for each GUI element. Let us see another example that may help to clarify the point:

```
1)  #include <GUIConstantsEx.au3>
2)  Opt("GUIOnEventMode", 1)
3)  $FADI = GUICreate("FakeDilutor", 200, 150, 500, 200)
4)  GUISetOnEvent($GUI_EVENT_CLOSE, "GUIExit")
5)  $Label1 = GUICtrlCreateLabel("Original solution volume (between
0.1 and 1):",20,20,140,25)
6)  $Input1 = GUICtrlCreateInput("",150,20,25,20)
7)  $Label2 = GUICtrlCreateLabel("Dilution water volume (between 0.1
and 10):",20,60,140,25)
8)  $Input2 = GUICtrlCreateInput("",150,60,25,20)
9)  $Button1 = GUICtrlCreateButton("Go", 50, 100, 50, 25)
10) GUICtrlSetOnEvent(-1, "CalculateDilutionOnEvent")
11) GUISetState()
12) While 1
13)     Sleep(500)
14) WEnd
15) Func GUIExit()
16)     Exit
17) EndFunc
18) Func CalculateDilutionOnEvent()
19)     If GUICtrlRead($Input1) >= 0.1 and GUICtrlRead($Input1) <= 1
and GUICtrlRead($Input2) >= 0.1 and GUICtrlRead($Input2) <= 10 Then
20)         $total = GUICtrlRead($Input1) + GUICtrlRead($Input2)
21)         $dilution = $total / GUICtrlRead($Input1)
22)         MsgBox(0,"Now diluting","Final volume:
"&$total&@CR&"Dilution: "&$dilution)
23)     Else
24)         MsgBox(16,"Error","Please input correct values")
25)     EndIf
26) EndFunc
```

Code sample 20.2 GUI with a functional button, equivalent to Code sample 18.7.

If you run Code sample 20.2 after removing the line numbers, you will see the same GUI that you saw when running Code sample 18.7. In fact, Code sample 20.2 was written with the purpose of showing an event mode-based GUI that performs the same as a message loop GUI. By analyzing this code, you will see the scheme that was outlined in the previous paragraph.

Comparing Code sample 20.2 and Code sample 18.7, the first impression is the longer code: 25 versus 17 lines. It is not a huge difference, but this, together with the need of thinking in another framework, may discourage some users to adopt event mode. In fact, so far, there seems to be no reason to do that. So, what would be a reason to write a code in event mode, and not in message loop mode?

20.2 Multitasking Using GUIOnEventMode

The main reason to use event mode instead of message loop mode, at least in the context of laboratory automation, is that using event mode it is much easier to enable "multitasking." I put multitasking between quotation marks because, technically, AutoIt is not capable of multitasking (explaining why is beyond the scope of this book). However, for simple purposes, we can adopt this term. Let us see an example that illustrates the point:

```
#include <GUIConstantsEx.au3>
GUICreate("Unstoppable", 200, 150)
GUISetState()
While 1
    For $i = 1 To 30
        ConsoleWrite($i)
        Sleep(100)
    Next
    If GUIGetMsg() = $GUI_EVENT_CLOSE Then Exit
WEnd
```

Code sample 20.3 GUI based on message loop mode that cannot be stopped.

```
#include <GUIConstantsEx.au3>
Opt("GUIOnEventMode",1)
GUICreate("Stoppable", 200, 100, 500, 200)
GUISetOnEvent($GUI_EVENT_CLOSE, "CloseButton")
GUISetState()
while(1)
    For $i = 1 To 30
        ConsoleWrite($i)
        Sleep(100)
    Next
WEnd
Func CloseButton()
    Exit
EndFunc
```

Code sample 20.4 GUI similar to that in Code sample 20.3, but based on event mode and that can be stopped.

First, run Code sample 20.3. You should see the console (the area on SciTE where some actions of the script are shown, see description in Section 2.5.2) displaying the numbers from 1 to 30 and starting repeatedly. This was done by the function "ConsoleWrite," which was called inside the infinite loop. Try to close the GUI, by pressing the close button, and nothing will happen. The only way to stop the script is by going to SciTE and using the Tools menu and calling Stop Executing.

Now, run Code sample 20.4. You should, again, see the console being filled with numbers from 1 to 30, repeatedly. However, this time, if you press the close button, the GUI finishes and the writing stops.

These two examples illustrate that in some situations the message loop mode is not adequate, and that the event mode needs to be used instead. Let us see an example that demonstrates how event mode can be useful in a laboratory automation context. The two codes below are for the same balance that was used as an example in Chapter 19. You need to have access to it in order to fully test the codes.

```
1)  #include "commMG.au3"
2)  #include <FileConstants.au3>
3)  #include <GUIConstantsEx.au3>
4)  #include <ColorConstants.au3>
5)  #include <StaticConstants.au3>
6)  Local $BalancePort = 0
7)  Local $PortError
8)  $StoringFile = "C:\balance\BalanceReading.txt"
9)  $ToledoBalance=GUICreate("Balance controller", 200, 160, 500, 200)
10) GUICtrlCreateTab(10, 10, 180, 140)
11) GUICtrlCreateTabItem("COM port")
12) $COMButton = GUICtrlCreateButton("Set COM port", 20, 40, 80, 20)
13) GUICtrlCreateTabItem("Weigh")
14) $Reading = GUICtrlCreateLabel("",80,40,80,18,$SS_RIGHT)
15) GUICtrlSetBkColor($reading,$COLOR_MEDGRAY)
16) $TareButton = GUICtrlCreateButton("Tare", 20, 60, 50, 20)
17) $WeighButton = GUICtrlCreateButton("Weigh", 20, 80, 50, 20)
18) $StoreButton = GUICtrlCreateButton("Store", 20, 100, 50, 20)
19) GUICtrlCreateTabItem("Options")
20) $ChangeButton=GUICtrlCreateButton("Change file", 20, 40, 80, 20)
21) GUISetState()
22) While 1
23)     If $BalancePort <> 0 Then ReadContinuous()
24)     Switch GUIGetMsg()
25)         Case $GUI_EVENT_CLOSE
26)             Exit
27)         Case $COMButton
28)             $BalancePort = InputBox("Set COM port","Type in COM port number, check Control Panel, Hardware settings.")
29)         Case $TareButton
30)             _CommSendStandard($BalancePort,$PortError,"T")
31)         Case $WeighButton
32)             _CommSendStandard($BalancePort,$PortError,"S")
33)             $BalanceReading = _CommReadString(50000)
34)             $ReadingArray = StringSplit($BalanceReading," ")
35)             GUICtrlSetData($Reading,$ReadingArray[4])
36)         Case $ChangeButton
37)         $StoringFile=FileSaveDialog("New file name","C:\","(*.txt)")
38)         Case $StoreButton
39)             $Name = InputBox("Store your data","The weight of this sample is " & $ReadingArray[4] & " . Add a sample name below:")
40)         FileWrite($StoringFile,$Name &" "& $ReadingArray[4] & @CRLF)
41)     EndSwitch
42) WEnd
43) Func ReadContinuous()
44)     _CommSendStandard($BalancePort,$PortError,"SI")
45)     $BalanceReading = _CommReadString(1000)
46)     $ReadingArray = StringSplit($BalanceReading," ")
47)     GUICtrlSetData($Reading,$ReadingArray[4])
48) EndFunc
```

Code sample 20.5 GUI based on message loop mode that controls the balance described in Chapter 19. This code is very similar to Code sample 19.5.

```autoit
1)  #include "commMG.au3"
2)  #include <FileConstants.au3>
3)  #include <GUIConstantsEx.au3>
4)  #include <ColorConstants.au3>
5)  #include <StaticConstants.au3>
6)  Opt('GUIOnEventMode', 1)
7)  Local $BalancePort = 0
8)  Local $PortError
9)  Local $ReadingForFile
10) $StoringFile = "C:\balance\BalanceReading.txt"
11) $ToledoBalance=GUICreate("Balance controller",200,160,500,200)
12) GUISetOnEvent($GUI_EVENT_CLOSE, "GUIExit")
13) GUICtrlCreateTab(10, 10, 180, 140)
14) GUICtrlCreateTabItem("COM port")
15) $COMButton = GUICtrlCreateButton("Set COM port", 20, 40, 80, 20)
16) GUICtrlSetOnEvent(-1, "SetCOMPortOnEvent")
17) GUICtrlCreateTabItem("Weigh")
18) $Reading = GUICtrlCreateLabel("",20,40,80,18,$SS_RIGHT)
19) GUICtrlSetBkColor($reading,$COLOR_MEDGRAY)
20) $TareButton = GUICtrlCreateButton("Tare", 20, 60, 50, 20)
21) GUICtrlSetOnEvent(-1, "TareOnEvent")
22) $WeighButton = GUICtrlCreateButton("Weigh", 20, 80, 50, 20)
23) GUICtrlSetOnEvent(-1, "WeighOnEvent")
24) $StoreButton = GUICtrlCreateButton("Store", 20, 100, 50, 20)
25) GUICtrlSetOnEvent(-1, "StoreOnEvent")
26) GUICtrlCreateTabItem("Options")
27) $ChangeButton=GUICtrlCreateButton("Change file", 20, 40, 80, 20)
28) GUICtrlSetOnEvent(-1, "ChangeFileOnEvent")
29) GUISetState()
30) While 1
31)     Sleep(500)
32)     If $BalancePort <> 0 Then ReadContinuous()
33) WEnd
34) Func ReadContinuous()
35)     _CommSendStandard($BalancePort,$PortError,"SI")
36)     $BalanceReading = _CommReadString(1000)
37)     $ReadingArray = StringSplit($BalanceReading," ")
38)     GUICtrlSetData($Reading,$ReadingArray[4])
39) EndFunc
40) Func GUIExit()
41)     Exit
42) EndFunc
43) Func SetCOMPortOnEvent()
44)     $BalancePort = InputBox("Set COM port","Type in COM port number, check Control Panel, Hardware settings.")
45) EndFunc
46) Func TareOnEvent()
47)     _CommSendStandard($BalancePort,$PortError,"T")
48) EndFunc
49) Func WeighOnEvent()
50)     _CommSendStandard($BalancePort,$PortError,"S")
51)     $BalanceReading = _CommReadString(50000)
52)     $ReadingArray = StringSplit($BalanceReading," ")
53)     GUICtrlSetData($Reading,$ReadingArray[4])
54)     $ReadingForFile = $ReadingArray[4]
55) EndFunc
56) Func StoreOnEvent()
57)     $Name = InputBox("Store your data","The weight of this sample is " & $ReadingForFile & ". Add a sample name below:")
58)     FileWrite($StoringFile,$Name & " " & $ReadingForFile & @CRLF)
59) EndFunc
60) Func ChangeFileOnEvent()
61)     $StoringFile = FileSaveDialog("Type new file name","C:\","(*.txt)")
62) EndFunc
```

Code sample 20.6 GUI based on event mode that controls the balance described in Chapter 19.

First, run Code sample 20.5. If you have the balance connected to your computer, you will see that it is very difficult to do anything with the GUI: the reading field keeps updating the reading and clicks on the buttons do not result in any effect. In fact, even closing the GUI is difficult. The best way to close the GUI, in this case, is to go to SciTE and tell the script to stop (Tools menu, then Stop Executing). If you do so, you may be surprised to see that the actions that you may have performed will happen after closing the script (at least some of them). This is clearly very unsatisfactory.

Now, run Code sample 20.6. You should see the reading field being updated constantly, and should be able to send commands to the balance, which are promptly executed. You should be able to exit the GUI by pressing the close button, too. In summary, it works very well. Therefore, it is clear that for this type of application, in which a task is constantly being performed and another needs to "jump in," it is necessary that the code be written in event mode.

20.3 Summary

- GUIs created in the previous chapters were limited to a single action at a given time. Using the GUIOnEventMode option enables multitasking.
- The GUIOnEventMode option works under a very different code structure compared to the one presented in the previous chapters about GUI.
- When using GUIOnEventMode, the function GuiGetMsg is not used any more. Instead, the function GuiSetOnEvent is called, and specific functions for each action are required.
- If you try multitasking in the message loop mode (using the function GuiGetMsg), your script will not work properly, while using GUIOnEventMode it will.
- In our example, using GUIOnEventMode enabled the continuous reading a balance output while operating the balance.

21

Adding Graphical Elements to a GUI

In addition to the typical and familiar controls like buttons and input fields, AutoIt also allows the insertion of nonstandard graphical elements in graphical user interfaces (GUIs). This can serve many purposes, from the purely aesthetic to the animated display of real-time measurements, which is a fairly common feature of analytical programs. AutoIt provides several different options to create graphical elements. In this chapter, only one of them, GDIplus is presented.

GDI stands for graphics device interface, and is a part of the Windows API (application programming interface). In other words, GDIplus is a set of standards that programmers can use to enable a task, in this case the drawing of graphical elements. There is a UDF (user-defined functions) library for AutoIt, which comes in the standard installation package.

I chose GDIplus, and not other options, for several reasons. One is what that has been already mentioned, that is, it comes readily available in the installation package. Another is that GDIplus works for all Windows versions from Windows XP on. GDIplus is also more powerful than the standard AutoIt graphical functions for GUIs (GUICtrl-CreateGraphic and GUICtrlSetGraphic; not examined in the book). Finally, the AutoIt community has worked extensively on GDIplus, and provided many examples of its usage online.

21.1 Getting Started with GDIplus

Using GDIplus, we can add simple graphical elements to a GUI:

```
#include <GUIConstantsEx.au3>
#include <GDIPlus.au3>

Dim $Graphs[5]
$SimplePlot = GUICreate("Shapes", 200, 150, 500, 200)

GUISetState()
_GDIPlus_Startup()
For $i = 0 to 4
    $Graphs[$i] = _GDIPlus_GraphicsCreateFromHWND($SimplePlot)
Next
```

21 Adding Graphical Elements to a GUI

```
$Pen1 = _GDIPlus_PenCreate(0xFFFF0000,3)
_GDIPlus_GraphicsDrawRect($Graphs[0],0,10,5,20,$Pen1)
$Brush1 = _GDIPlus_BrushCreateSolid(0xFFFFF000)
_GDIPlus_GraphicsFillRect($Graphs[1], 10, 10, 30, 30,$Brush1)
$Pen2 = _GDIPlus_PenCreate2($Brush1,4)
_GDIPlus_GraphicsDrawEllipse($Graphs[2],50,10,30,20,$Pen2)
_GDIPlus_GraphicsFillEllipse($Graphs[3],100,10,20,20)
_GDIPlus_PenSetDashStyle($Pen1,$GDIP_DASHSTYLEDASH)
_GDIPlus_GraphicsDrawLine($Graphs[4],20,60,180,60,$Pen1)
_GDIPlus_PenSetLineCap($Pen2,0x14,0x11,3)
_GDIPlus_GraphicsDrawLine($Graphs[4],30,80,150,100,$Pen2)

Do
Until GUIGetMsg() = $GUI_EVENT_CLOSE

_GDIPlus_PenDispose($Pen1)
_GDIPlus_PenDispose($Pen2)
_GDIPlus_BrushDispose($Brush1)
For $i = 0 to 4
   _GDIPlus_GraphicsDispose($Graphs[$i])
Next
_GDIPlus_Shutdown()
```

Code sample 21.1 Introducing some simple GDIplus graphical elements.

Run the script, and you should see a GUI like the one in Figure 21.1.

As you can see, different types of graphical elements can be found in the GUI. Let us examine the code in detail to fully understand it. Before getting to the GDIplus part, observe that the GUI endless loop was not of the While type, but a Do … Until loop. They work essentially in the same way. Here, the loop is simply a line with the "Do" instruction, and the next line already has the "Until" instruction. Normally, we would add some instructions between those two, but here the loop is called simply to keep the GUI open, as we learned in Chapter 18. The constant "$GUI_EVENT_CLOSE" corresponds to the close button. Therefore, the meaning of the Do … Until loop here is to repeat until the close button is clicked.

As you know by now, it is necessary to include the file referring to the functions to be called in the code, which for GDIplus is "GDIplus.au3." Then, it is necessary to call the function "_GDIPlusStartup," just after the GUISetState function. If you call _GDIPlusStartup, at the end of the code you must also call "_GDIPlusShutDown."

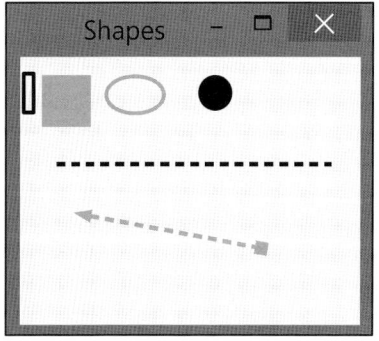

Figure 21.1 GUI created using Code sample 21.1.

The next function that must be called is "_GDIPlus_GraphicsCreateFromHWND," which receives as argument the GUI where the graphics will appear. This function must be passed to the graphics that will be drawn. Here, it was passed to an array, which has drawings as elements. It is not necessary to use an array in this part of the code, but this is convenient in order to avoid the repetition of the _GDIPlus_GraphicsCreateFromHWND instruction. If _GDIPlus_GraphicsCreateFromHWND is called, at the end of the code the function "_GDIPlus_GraphicsDispose" must also be called, receiving as arguments all graphic elements created using _GDIPlus_GraphicsCreateFromHWND.

After the creation of the graphical elements, we can start determining their attributes, like shape, location, and color. In our example, the first function being called is "_GDIPlus_PenCreate," which is passed to the variable "$Pen1" and receives as arguments a color (in the hexadecimal base) and its thickness (3). This means that any graphic drawn with this pen will have this color and line thickness. For example, the next instruction, "_GDIPlus_GraphicsDrawRect," receives $Pen1 as its last argument. It also receives Graph[0], that is, one of the graphics created using _GDIPlus_GraphicsCreateFromHWND, and values for its position and size. In the next line, a brush is created using "_GDIPlus_BrushCreateSolid," which is used for the next graphical element, a filled rectangle. Filled elements need to be painted using a brush, and not a pen. Another pen ("$Pen2") is created receiving the parameters from the brush, and is used to draw the following ellipse. After this ellipse, another ellipse is drawn but receiving neither pen nor brush, meaning that it will be black and drawn using the finest pen (if applicable). The next graphical element is a line drawn using $Pen1, and the last one is another line drawn using $Pen2, but after it has been modified to include line caps of two different types. Similarly to graphical elements, pens and brushes must be discarded at the end of the code.

There are many more options for drawing graphics using GDIplus, with fairly advanced graphical effects being available. Interested readers can explore the AutoIt help file and experiment with the many different functions if they intend to create more advanced graphical elements.

21.2 Creating Animations Using GDIplus

A very useful feature that can be enabled using GDIplus is the creation of animations. Such animations can be used to create chromatograms and other common analytical outputs often available in modern laboratory software. As an example, let us enable real-time measurement of the output of the Restek ProFlow 6000 flowmeter, as described in Section 17.3. As there, let us start by reproducing the values of the computer clock:

```
#include <GUIConstantsEx.au3>
#include <GDIPlus.au3>

Opt("GUIOnEventMode",1)

Dim $Rects[30]
Dim $x[30]
Dim $y[30]
Dim $oldy[30]
```

21 Adding Graphical Elements to a GUI

```
$SimplePlot = GUICreate("Moving shapes", 350, 150, 500, 200)
GUISetOnEvent($GUI_EVENT_CLOSE, "CloseButton")
GUISetState()
_GDIPlus_Startup()
$MyBrush = _GDIPlus_BrushCreateSolid()
_GDIPlus_BrushSetSolidColor($MyBrush, 0xFFFF0000)
For $i = 0 to 29
    $Rects[$i] = _GDIPlus_GraphicsCreateFromHWND($SimplePlot)
    $x[$i] = 10 + 10*$i
    $y[$i] = 100
    $oldy[$i] = 100
Next

While 1
    For $i = 29 to 1 Step -1
        _GDIPlus_GraphicsFillRect($Rects[$i], $x[$i], $y[$i], 10, 2,$MyBrush)
        $y[$i-1] = $oldy[$i]
        $oldy[$i] = $y[$i]
        $y[$i] = 100-@SEC
    Next
    Sleep(1000)
    _GDIPlus_GraphicsClear($Rects[$i],0xFFFFFFFF)
WEnd

_GDIPlus_BrushDispose($MyBrush)
_GDIPlus_GraphicsDispose($Rects)
_GDIPlus_Shutdown()

Func CloseButton()
    Exit
EndFunc
```

Code sample 21.2 Script that produces a simple animation describing the change in the seconds as received from the computer clock.

Run the script, and you should see a line composed of rectangles at the bottom of the window, and then coming up starting from the right-hand side (snapshot in Figure 21.2).

Before analyzing the code, let us understand how animation works. The illusion of animation is enabled by drawing the graphic, then erasing it, and then redrawing it at a new, slightly different, position. If the three steps are done fast enough, people perceive

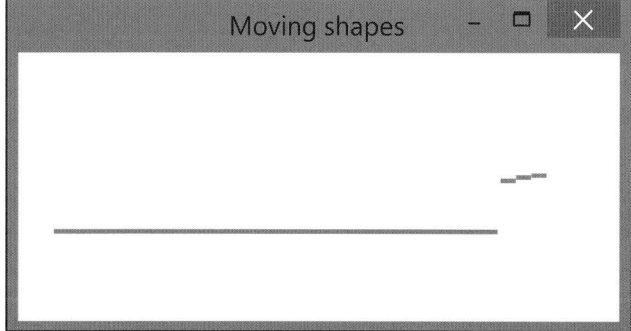

Figure 21.2 GUI that responds to the windows clock by drawing rectangles at different positions, see Code sample 21.2.

them as the graphic moving. Therefore, in addition to functions that draw graphics, we need a function to erase the drawings in the code.

Code sample 21.2 was written using the OnEventMode option (see full details in Chapter 20), which was necessary to enable the access to the closing button on the window. The GDIplus functions being called follow the general pattern shown in Code Sample 21.1, the exception being the function "_GDIPlus_GraphicsClear." This function erases the graphics passed as argument (in this case, the array "Rects" and substitute them with a background color). Note that the background color was different from the initial one, which I did on purpose to emphasize this step.

The other arrays in the code ("$x," "$y," and "$oldy") are passed as the coordinates for the rectangles. Because the idea is to have a display that works like the Calc plots being updated (Section 17.3) from right to left, each rectangle is initially assigned with $y and $oldy having the same value. This value is then changed for $y after the older $y value has been passed to $oldy and the next rectangle receives the $oldy value from the rectangle being updated. The loop continues indefinitely.

Now we can adapt the code for the flowmeter:

```
#include "commMG.au3"
#include <GUIConstants.au3>
#include <GDIPlus.au3>

Opt("GUIOnEventMode",1)

Local $port = 4
Local $portError

Dim $HLabels[16]
Dim $VLabels[11]
Dim $HLines[11]
Dim $Rects[30]
Dim $x[30]
Dim $y[30]
Dim $oldy[30]

_CommSetPort($port, $portError, 115200, 8, 0, 1, 2)

$FlowMeter = GUICreate("Flow Meter Monitor", 500, 650, 10, 10)
GUISetState(@SW_SHOW, $FlowMeter)
GUISetOnEvent($GUI_EVENT_CLOSE, "CloseButton")
$Label1 = GUICtrlCreateLabel("mL / min",240,20,100,30)
GUICtrlSetFont($Label1,12,700)
For $i = 0 to 15
    $HLabels[$i]=GUICtrlCreateLabel($i,410-$i*20,560,15,20)
Next

$Label2 = GUICtrlCreateLabel("Elapsed time (s)",235,600,100,20)

$SimplePlot = GUICreate("Plot", 350, 550, 100, 50, $WS_CHILD, -1, $FlowMeter)
GUISetState(@SW_SHOW, $SimplePlot)
_GDIPlus_Startup()
$BigRect = _GDIPlus_GraphicsCreateFromHWND($SimplePlot)
For $i = 0 to 10
    $VLabels[$i] = _GDIPlus_GraphicsCreateFromHWND($SimplePlot)
    $HLines[$i] = _GDIPlus_GraphicsCreateFromHWND($SimplePlot)
Next
$MyBrush = _GDIPlus_BrushCreateSolid()
_GDIPlus_BrushSetSolidColor($MyBrush, 0xFF0000FF)
For $i = 0 to 29
    $Rects[$i] = _GDIPlus_GraphicsCreateFromHWND($SimplePlot)
    $x[$i] = 10 + 10*$i
    $y[$i] = 499
    $oldy[$i] = 499
Next
```

21 Adding Graphical Elements to a GUI

```
$step = 0
While 1
   $flow = ReadFlowMeter(1000)
   If $step <> 0 Then _GDIPlus_GraphicsClear($Rects[$i],0xFFF0F0F0)
   For $i = 29 to 1 Step -1
      _GDIPlus_GraphicsFillRect($Rects[$i], $x[$i], $y[$i], 10, 2,$MyBrush)
      $y[$i-1] = $oldy[$i]
      $oldy[$i] = $y[$i]
      $y[$i] = 499-$flow
      $step = 1
   Next
   _GDIPlus_GraphicsDrawRect($BigRect,20,0,290,500)
   For $i = 0 to 10
      _GDIPlus_GraphicsDrawLine($HLines[$i],20,50*($i),310,50*($i))
      _GDIPlus_GraphicsDrawString($VLabels[$i],50*($i),0,50*(10-$i),"Arial",8)
   Next
   Sleep(100)
WEnd

_GDIPlus_BrushDispose($MyBrush)
_GDIPlus_GraphicsDispose($BigRect)
For $i = 0 to 10
   _GDIPlus_GraphicsDispose($VLabels[$i])
Next
For $i = 0 to 10
   _GDIPlus_GraphicsDispose($HLines[$i])
Next
For $i = 0 to 29
   _GDIPlus_GraphicsDispose($Rects[$i])
Next
_GDIPlus_Shutdown()

Func CloseButton()
    Exit
EndFunc
```

Code sample 21.3 Script that displays a simple animation of the input from the Restek ProFlow 6000 flowmeter.

Run the script and, if you have the flowmeter connected to your computer at the correct COM port, you should see an output as that in Figure 21.3, which resembles the result obtained using the techniques described in Section 17.3, more specifically Code sample 17.6.

Code sample 21.3 is long for the standards of this book, but most of its elements should be familiar to readers following the book to this point. These are the different sections of the code (separated by empty lines in the code): (i) the inclusion of the necessary libraries; (ii) the setup of the OnEventMode option; (iii) the declaration of some variables (relative to the COM port being accessed, and some arrays); (iv) the setup of the COM port (function _CommSetPort); (v) the creation of the "$Flowmeter" GUI, including its elements; (vi) the creation of another GUI ("$SimplePlot"), including its elements and the start of the GDIplus definitions; (vii) a While loop where the real action happens, that is, the COM port is read, and the GDIplus graphical elements are drawn according to this information, and subsequently erased to be redrawn; (viii) the termination of GDIplus, with the disposal of graphical elements and the shutdown; and (ix) the definition of the function that exits the GUI using the close button. Among the several parts of the code, the only new element is the existence of a GUI ($SimplePlot) inside another GUI ($Flowmeter). This technique allows that only $SimplePlot is updated with the animation. Using this technique, you can create a complex GUI that displays animated graphics based on GDIplus, and have other controls in the other parts of the GUI. As a final note, observe that differently from the other GUI codes through the

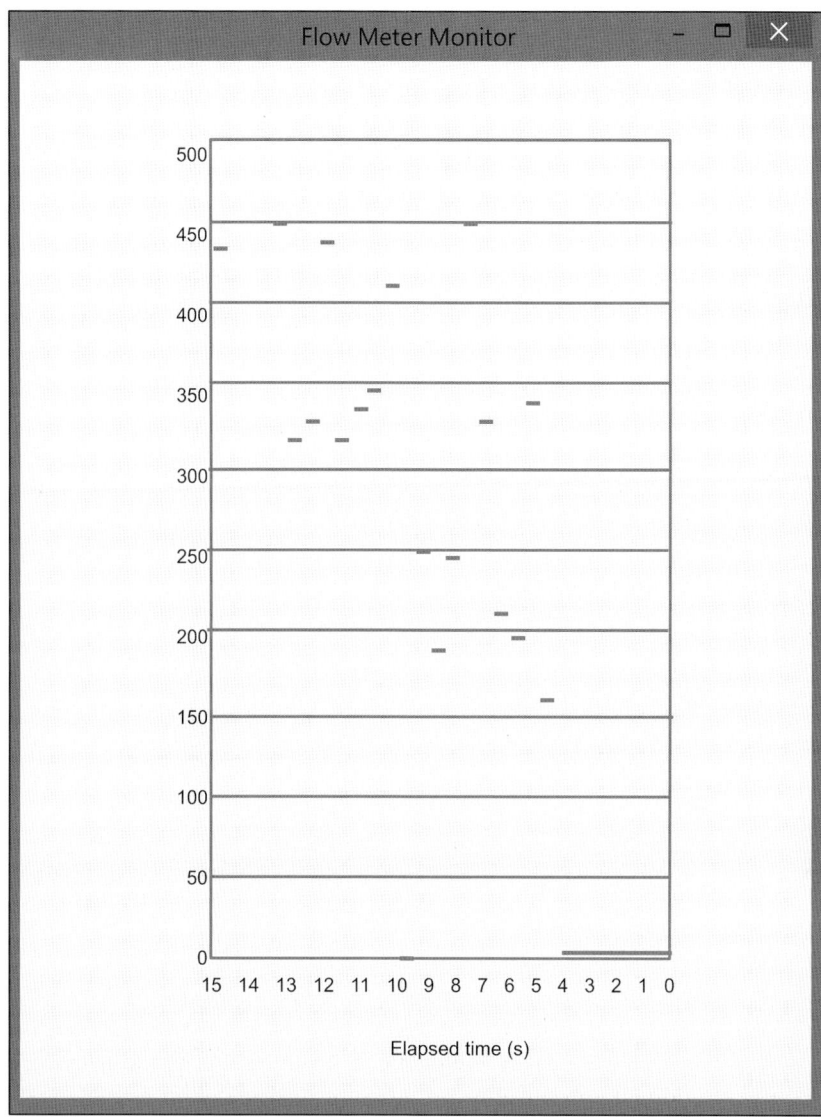

Figure 21.3 Display of the input from the Restek ProFlow 6000 flowmeter obtained using the script in Code sample 21.3.

book, GUIconstants.au3, and not GUIconstantsEx.au3 was included at the start of the code, which was necessary to enable the use of the constant "$WS_CHILD" when calling the function GUICreate.

21.3 Summary

- AutoIt allows the insertion of graphical elements in GUIs using the GDIplus technology.

- GDIplus is enabled by inserting the functions from the file GDIplus.au3.
- When using GDIplus, it is essential to call the functions _GDIplusStartup and _GDIPlus_GraphicsCreateFromHWND.
- The properties of graphical elements can be defined using many different functions.
- At the end of the code, the graphical elements must be discarded using the appropriate functions, and GDIplus must be shut down using the _GDIplusShutdown function.
- Animations are enabled by drawing the graphics, showing them for a short time, erasing them using the function _GDIPlus_GraphicsClear, and drawing them again. The process can be repeated indefinitely, generating the illusion of animation.
- Alternatively to While loops, Do … Until loops can be used to enable infinite loops when showing GUIs.
- A GUI can be called inside another GUI if convenient. In this case, the constant $WS_CHILD must be passed to the function GUICreate.

22

Creating GUIs Using Koda

Chapters 18–21 focused on graphical user interface (GUI) creation. As has been explained in Chapter 18, there is a tool named Koda that helps with GUI creation, but I have deliberately ignored it so far, because in order to grasp the basic concepts of GUI creation, it is easier if you write the code yourself. Now that you have gathered some background in GUI creation, it is useful to learn about Koda, as this tool makes it much easier to create complex GUIs.

22.1 Getting Started with Koda

In order to start Koda, you need to go to SciTE and save a script, even if it has nothing written on it. By doing so, you enable many tools in the Tools menu that are not available when there is no saved script open on SciTE. Once this is done, go to the tools menu and click on "Koda (FormDesigner)," or press Alt + m. By doing so, several windows will appear (Figure 22.1), and SciTE will become frozen in the same way as it becomes when you run a script. This, however, does not matter, as all the work will be done using the Koda interface without needing to access SciTE.

The main feature in Koda is that you can build a GUI in the style "what you see is what you get" (WYSIWYG). When you start Koda, the center of the attention is Form1 (Figure 22.1). You can change Form1 size by dragging on its edges, in the same way you can do with any window. Therefore, instead of using GUICreate to determine the dimensions of your GUI, as done in the previous chapters, you can simply draw and shape it. How about changing its title, or making sure of its size, and so on? For this, you can use the "Object Inspector," at the left of Form1 in Figure 22.1. With that tool, you can define all relevant parameters of the "object" (I put within " ", because the term is not necessarily synonymous with the technical definition of objects in programming) being edited. For Form1, for example, you can change its "Caption," which is the same as the title, for a Window. You can go ahead and explore the many options that are available.

Next, you can go to the task bar just above Form1 in Figure 22.1. You will find there the familiar controls that were used in previous chapters, and some others too. To add one of such controls to Form1, you simply click on the icon and then move the mouse pointer to the Form1 area, and click-dragging the mouse pointer to the end position. This way, you can directly set the position and size of the control (as you should probably know by now, control is the technical name of the elements of a GUI; mode details in Chapter 5). You can modify it anytime by clicking at its borders and resizing it, or by clicking on

Practical Laboratory Automation: Made Easy with AutoIt, First Edition. Matheus C. Carvalho.
© 2017 Wiley-VCH Verlag GmbH & Co. KGaA. Published 2017 by Wiley-VCH Verlag GmbH & Co. KGaA.

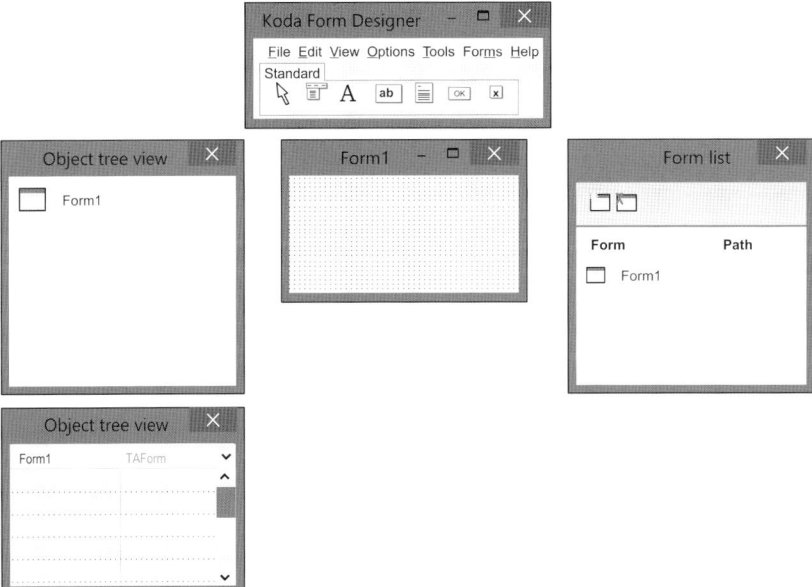

Figure 22.1 Main elements of the Koda interface.

it and dragging it to a different position. Similarly to the whole form, you can do finer adjustments to it by using the Object Inspector. You could try, for example, adding a button. Select the button (by clicking on it) and then go to the Object Inspector. Note that the list of options is different for a button and a form. As for the form, you will be able to set many formatting parameters, including some advanced ones, like setting a picture on it, for example.

In addition to helping with formatting single controls, Koda makes the task of grouping and organizing GUI controls very easy. For example, add another button to your form. Click and drag the left mouse button creating a rectangle around the two buttons, and then click on the selected area with the right button. A menu will appear (Figure 22.2) offering some options to be selected.

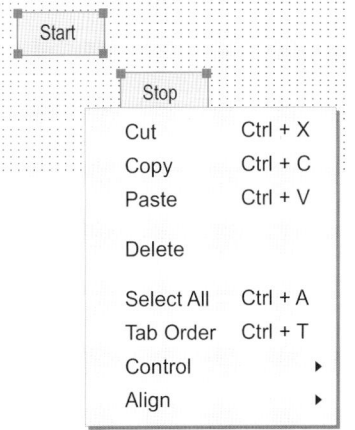

Figure 22.2 If you select one or more controls in a form, and click on them using the right button, a menu appears.

Figure 22.3 Submenu activated when Align is chosen in the menu in Figure 22.2.

Left
Center
Right

Top
Middle
Bottom

Space Equally Horizontally
Space Equally Vertically

Center Horizontally in Window
Center Vertically in Window

Equal width
Equal Height
Equal Size

Align to grid

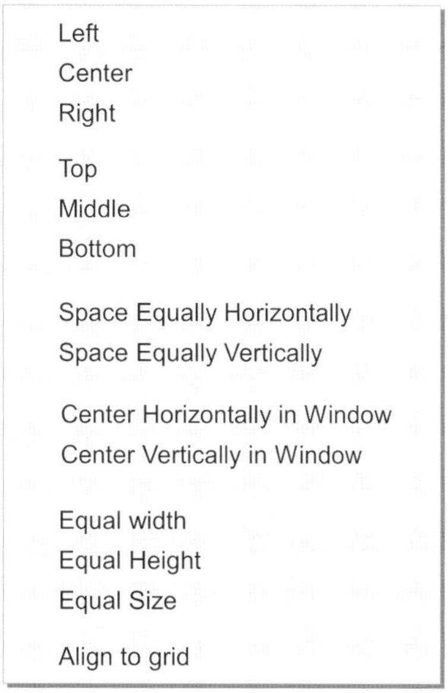

Figure 22.4 Submenu activated when Control is chosen in the menu in Figure 22.2.

Bring to front
Send to back

Lock
Unlock
No insert in
No copy

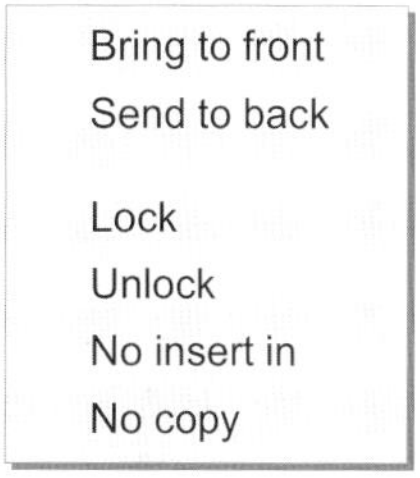

Note that the menu offers familiar operations like cut, copy, paste, and delete. In addition, there are many other options that can help you to create a personalized GUI. A very useful one is "Align," which allows you to not only organize the controls according to a fixed position, but also resize the selected controls (Figure 22.3).

Another option in the menu in Figure 22.2 is "Control." By clicking on it, you open another submenu (Figure 22.4).

Bringing a control to front means that, if you place it at the same position of another control, the former will be above and the latter below. The opposite is true when you choose to send the control to back. Locking a control means that you cannot move or edit it. For example, create several buttons and lock one of them. Then, select all of them and try to move or resize them. All but the locked button will be moved or resized. "No insert in" is an option that applies to only some types of controls, like groups and tabs, and means that you cannot insert a new control in one of them. Finally, "No copy" means that you cannot copy and paste the control. Finally, you can try "Tab Order." The purpose of this option is to determine how the focus will change from control to control as you press the TAB key when the script is running.

If you try to save your GUI (Ctrl + s or using the File menu), you will see that the type of file (.kxf) is not the same as for a normal AutoIt script file (.au3). This is because Koda saves a "project," and not a script. In the next section, we will see how to convert a project to a script. If you save your project, you can open it later. Also, you can have more than one project open at the same time.

The other two small windows completing the Koda interface are "Object TreeView" and "Form List" (see Figure 22.1). The first is useful to help organize the controls in a form, and the latter to help organize several forms if open at the same time. Therefore, both are mostly useful for larger projects.

22.2 Creating a Script

Once you have some controls in your Form, you need them to do something useful, as explained in Chapter 18. Koda generates code automatically for the controls in the form. You can test this by pressing F10, which will show the code in SciTE (Figure 22.5).

Looking at the code in Figure 22.5, the less familiar element is "#region," which, by its color, you can guess that does not belong to the workable part of the code. In fact, #region is a type of comment, which states that a certain part of the code starts at #region and ends at #EndRegion. Between these two delimiters, we see that the Form and the buttons are all declared, already with their labels, dimensions, and positions. Also, the function GUISetState is called with the @SW_SHOW macro as an argument. This argument is optional, as we saw in Chapters 19–21 that GUISetState can be called without any argument. The remaining of the code is the While loop of the standard GUI mode containing the function GUIGetMsg.

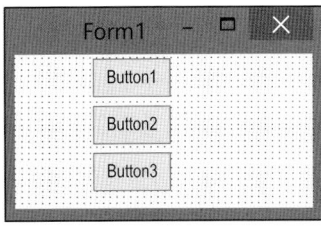

```
#include <ButtonConstants.au3>
#include <GUIConstantsEx.au3>
#include <WindowsConstants.au3>
#Region ### START Koda GUI section ### Form=c:\prettypics\form1.kxf
$Form1 = GUICreate("Form1", 615, 437, 192, 124)
$Button1 = GUICtrlCreateButton("Button1", 144, 40, 105, 33)
$Button2 = GUICtrlCreateButton("Button2", 144, 88, 105, 33)
$Button3 = GUICtrlCreateButton("Button3", 144, 128, 105, 33)
GUISetState(@SW_SHOW)
#EndRegion ### END Koda GUI section ###

While 1
    $nMsg = GUIGetMsg()
    Switch $nMsg
        Case $GUI_EVENT_CLOSE
            Exit

        Case $Button1
    EndSwitch
WEnd
```

Figure 22.5 Example of code generated from a Form in Koda.

One thing that you may find strange is that only $Button1 is being called inside the loop, and the other two variables referent to the other buttons, $Button2 and $Button3, are not. This is happening because, before generating the code, I told Koda to attach $Button1 to an event. I did that by clicking twice on Button1 in Form1 in the Koda interface. If you do so with any button, you will see a window like that shown in Figure 22.6.

On the "Event Editor" window, you can choose the "Control notification" to be "None" or "Notify." If you choose Notify, then the control will be included in the event list in the While loop when the code is generated. Control notification is none by default, and therefore no event is linked to the control in the While loop unless you specify it using the Event Editor.

As you know by now, the code in Figure 22.5 was generated in the message loop mode, which is the default GUI mode for AutoIt. It is also possible to generate code in OnEvent-Mode using Koda. In order to enable that, you need to choose Options, which you can do by either using the menu or pressing Ctrl + k (Figure 22.7).

There, you select "Code Generator" and then activate "Generate OnEvent Code." Doing so, the code that you generate by pressing F10 will be on OnEventMode.

There is much more that Koda can do, and interested readers are encouraged to explore the features by using the program. If you followed the GUI chapters so far, the help file should be enough for you to make progress in the subject, and decide whether Koda or manual code writing is the best option for your project.

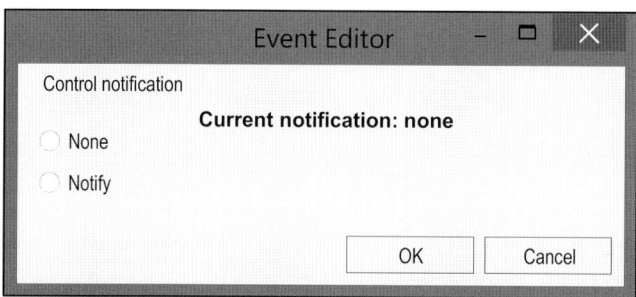

Figure 22.6 Setting an event to be linked to a control.

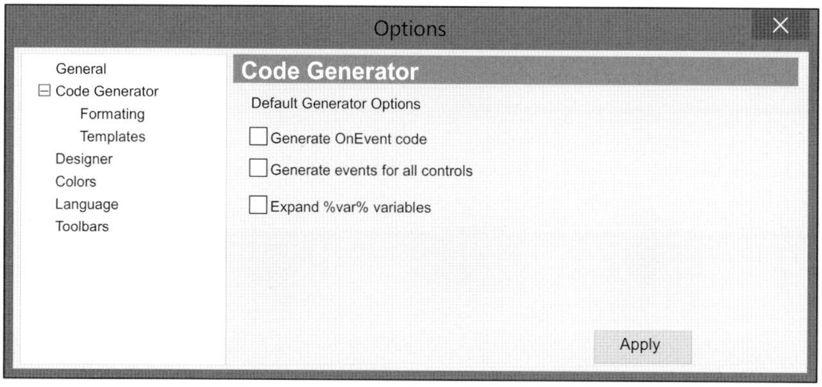

Figure 22.7 Choosing the generated code to be in OnEventMode.

22.3 Summary

- Koda is a graphical editor for GUI creation in AutoIt.
- Once you have some familiarity with GUI functions, Koda can be very useful.
- You can directly reshape and resize GUI elements using Koda, and can do finer adjustments using the Object Inspector.
- Koda saves its work as projects, and not as scripts.
- Koda automatically generates the AutoIt code for the GUI.
- The generated code is by default in message loop mode, but you can choose to change it to on event mode.

23

Some Suggestions

In this chapter, my intention is to offer unsolicited advice to instrument users and manufacturers. By adopting the ideas presented here, laboratory automation can become more accessible for the average laboratory technician or scientist.

23.1 For Manufacturers: All Instruments with a GUI

Although it is possible to control instruments provided their communication protocols are made transparent, as it was explained in Chapter 17 for serial communications, it is much easier if a graphical user interface (GUI) is available to control the instrument. As seen in Chapters 18–22, to create a GUI is not difficult in many cases. If for a given instrument it is, then it is even clearer that a GUI must be made available to the user.

23.2 For Manufacturers: All GUIs with Access to Controls

It was seen in Chapter 5 that scripting with controls enables multitasking, in a really fluid manner so that it becomes seamless that a script is operating and synchronizing instruments. However, there are many cases in which controls are not transparent for AutoIt. The way to make controls transparent is to adopt the Windows API as a standard. The Windows API is fully described at https://msdn.microsoft.com/en-au/library/windows/desktop/ff818516(v=vs.85).aspx..

Note that manufacturers are not blocking AutoIt by not following this instruction; they are merely making it a little difficult for users to use scripting to automate your program. If it is too difficult or too expensive to implement a change of the kind, it is not necessary.

23.3 For Manufacturers: Stop Developing Standards for Laboratory Automation

There have been repeated efforts by sectors of the industry with the aim of adopting standards to facilitate communication between instruments from different manufacturers (full discussion in Chapter 1). I believe this book demonstrates that it is not necessary

Practical Laboratory Automation: Made Easy with AutoIt, First Edition. Matheus C. Carvalho.
© 2017 Wiley-VCH Verlag GmbH & Co. KGaA. Published 2017 by Wiley-VCH Verlag GmbH & Co. KGaA.

to engage in such endeavor. Adopting standards is a difficult task, which demands teams of professionals working to tackle the many difficult challenges involved in the process. This effort would be better directed to the development of new automation solutions or new measuring instruments. For example, fields like clinical microbiology lag in terms of automation.

Note that what is argued here does not necessarily contradict the previous recommendation, in which I suggest that the Windows API is adopted when developing new software. As I made clear there, I only suggest this if it will not be a burden in the current production process. In other words, if it is to implement standardization, then be that of adopting Microsoft standards for program writing (the Windows API).

Furthermore, I am only referring to software here. Efforts to standardize hardware are more than welcome, of course.

23.4 For Users: Hardware Trumps Software

Because AutoIt enables the improvement of the programs available to control an instrument, any argument regarding the superiority of a given controlling program above other becomes less relevant. For example, by comparing the default autosampler of a certain instrument with a low-cost 3D printer, if you conclude that the 3D printer can be used for sampling without problems, it matters very little that the program controlling the 3D printer was designed with 3D printing in mind, whereas that one for the purposely designed autosampler was designed for sampling.

23.5 For Users: If You Can, Choose Controls

As explained above for manufacturers, scripting with access to controls offers advantages compared to mouse- and keyboard-based scripting. Therefore, if you have the option to choose a program to control a device, or to export data, or for any other relevant task, you will probably be better of if you choose one that makes controls transparent for AutoIt.

23.6 For Users: AutoIt May Not be the Best Programming Option in Some Cases

In my opinion, AutoIt is the best option for the purposes outlined in this book, which are mostly concerned with integrating existing pieces of software to automate tasks in a laboratory. However, if your intention is to build a new piece of equipment and its software from scratch, then you will need to learn much in terms of programming, in addition to dealing with electronics and mechanics. For that purpose, other programming languages can be more adequate, like Visual Basic and LabView. There are many resources to learn such languages. Also, you may wish to learn about Arduino, Raspberry Pi, or other microcontrollers, in addition to 3D printer-based fabrication, that makes it easier to develop electronic hardware. See the list of bibliography in Chapter 7 about open-source hardware as a starting point.

23.7 For Users: Be Aware of Technological Advances

The current pace of technological development is extremely fast, to the point of being almost incomprehensible. Almost daily, mind-blowing new devices or pieces of software are created, some with revolutionary implications. With AutoIt, applying new and revolutionary devices to laboratory routines became easier than ever (check the comments about robotic arms in Chapter 7, for instance). The scientist/technician that knows how to use AutoIt, and keeps an eye open for these technological advances, will be ahead of the game.

23.8 For Users and Manufacturers: AutoIt Scripts May Serve as Basis for New Products

Because they are so easy to write, AutoIt scripts can be developed by technicians and scientists in laboratories. This way, these professionals can find solutions for their problems that are not available by default in the software provided by manufacturers of instruments, which are most times developed by professional programmers. Therefore, users can present these scripts to manufacturers in order to suggest future improvements in their products. Or, alternatively, manufacturers can search for user-made solutions as inspirations for their new products. It is a win–win game.

Suggested Reading

Butler, D. (2016) Tomorrow's world. *Nature*, **530**, 399–401.
Pearce, J.M. (2014) *Open-Source Lab: How to Build Your Own Hardware and Reduce Research Costs*, Elsevier.

Appendix A

Other SciTE Features

In Chapter 2, I introduced SciTE, the standard editor of AutoIt scripts, but I only presented its most common utilities. I decided to do so because SciTe has an extensive list of tools (Figure A.1), and most of them are not immediately useful when writing simple scripts. However, some useful SciTE features were not covered. This appendix aims to present some of them.

A.1 Code Wizard

Code Wizard is a tool that aims to make it easier to produce the code for some of the most common functions in AutoIt, including Message Box and Input Box, which were used through the book. You find it in the Tools menu on SciTE, or you press Alt + W while using SciTE, so that Code Wizard window opens (Figure A.2).

Code Wizard has seven tabs. The first one is for Message Boxes. In this book, I did not place much importance on Message Boxes, using them mainly as warnings. However, Message Boxes can serve a number of purposes, like offering options (Ok or Cancel, Yes or No, etc.). If you want to explore the full potential of Message Boxes, Code Wizard makes the task easier. You can combine the several different attributes for Message Boxes and press "Preview" to see what you get. If you are satisfied, then you can press "Copy 2 SciTE" and then the code for your message box will be incorporated into your script.

The next four tabs, "Input Box," "Tool Tip," "Splash Text," and "Splash Image," are quite similar among them. Except for Input Box, these functions were not covered in this book, as I found no use for them in the scripts. The options in Code wizard for them are quite straightforward, and you will easily understand if you try them.

The next two tabs help with GUI creation. "Colors" has a nice feature of naming the colors according to their hexadecimal values. These tools complement Koda (see Chapter 21), which is the full-featured tool for GUI creation for AutoIt.

Practical Laboratory Automation: Made Easy with AutoIt, First Edition. Matheus C. Carvalho.
© 2017 Wiley-VCH Verlag GmbH & Co. KGaA. Published 2017 by Wiley-VCH Verlag GmbH & Co. KGaA.

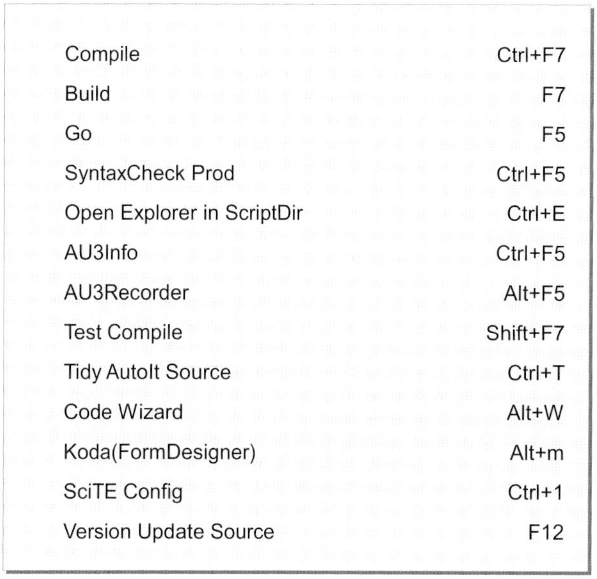

Compile	Ctrl+F7
Build	F7
Go	F5
SyntaxCheck Prod	Ctrl+F5
Open Explorer in ScriptDir	Ctrl+E
AU3Info	Ctrl+F5
AU3Recorder	Alt+F5
Test Compile	Shift+F7
Tidy AutoIt Source	Ctrl+T
Code Wizard	Alt+W
Koda(FormDesigner)	Alt+m
SciTE Config	Ctrl+1
Version Update Source	F12

Figure A.1 A portion of the long Tools menu on SciTE.

A.2 Organizing Your Scripts with Tidy

Tidy is a tool which organizes the code in your script based on some arbitrary conventions. Although arbitrary, they are pretty common in programming and thus make your script easier to be understood by others, and even by yourself, probably, after some time away from it. Let us compare how a script looks before and after Tidy:

```
Func FACACOpixel()
$Wcolor = 65280
$Pcolor = PixelGetColor(252,52)
While $Wcolor <> $Pcolor
sleep(20)
$Pcolor = PixelGetColor(252,52)
WEnd
EndFunc
```

Code sample A.1 Code manually typed without following any convention.

```
Func FACACOpixel()
   $Wcolor = 65280
   $Pcolor = PixelGetColor(252, 52)
   While $Wcolor <> $Pcolor
      Sleep(20)
      $Pcolor = PixelGetColor(252, 52)
   WEnd
EndFunc   ;==>FACACOpixel
```

Code sample A.2 The same code of Code sample A.1 after using Tidy.

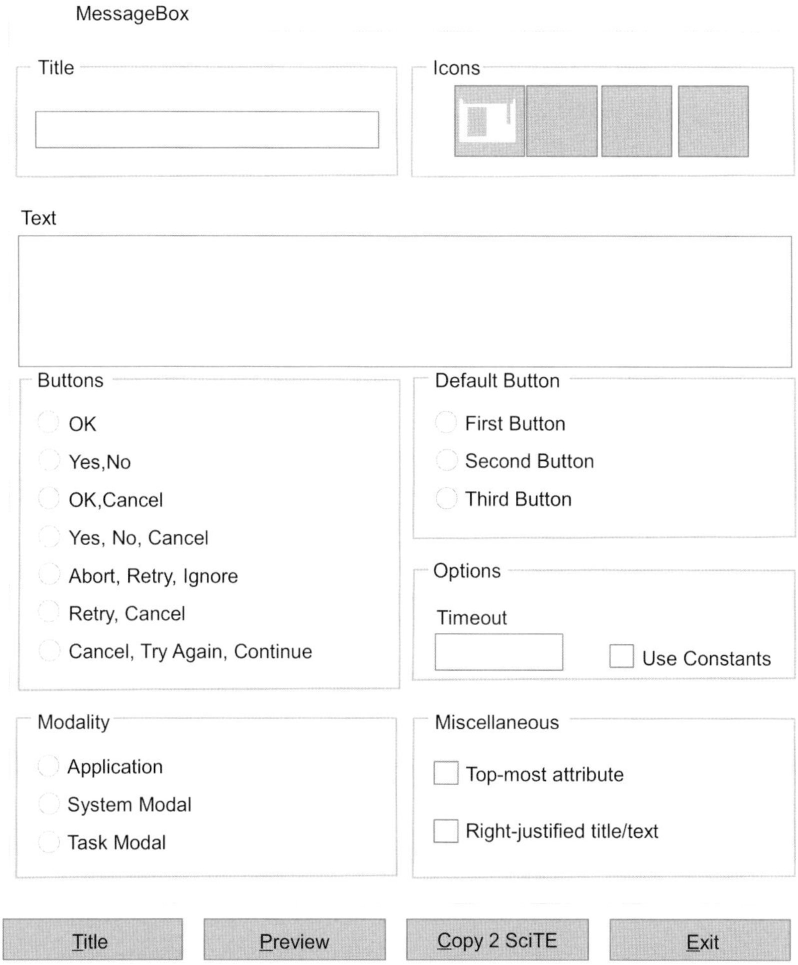

Figure A.2 MessageBox TAB on Code Wizard interface.

Code sample A.2 is arguably much easier to read than Code sample A.1. Even a comment was automatically generated to make it very clear that the function ended there.

Tidy can be activated on the Tools menu of SciTE or, more easily, by pressing Ctrl + T. It will correct the full code of the script. It will also create a backup of your old "untidy" code in a folder named "Backup" inside the folder you are working. However, by pressing Ctrl + Z once you can easily undo what Tidy did.

A.3 Tools that Facilitate Navigation

The scripts in this book were all short, some of them very short. The exception was the FACACOFAKASfunctions.au3 file, which, if you followed all the exercises in

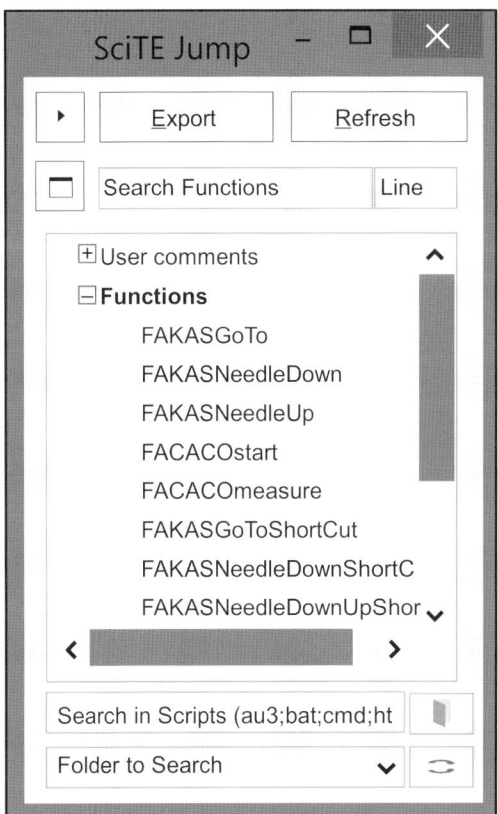

Figure A.3 SciTE jump showing some of the functions in the FACACOFAKASfunctions.au3 file.

the book, must have become enormous. Also, although I did the job for you, if you tried to study the UDFs that based COM port and IRC scripts, you saw that the codes were quite long. SciTE has some tools to make it easier to navigate such long codes.

A very useful tool to help navigation is SciTE jump (Alt + Q, Figure A.3). With SciTE jump, you can browse through the functions in the code. When you select one of them, SciTE will show it on its window. As you can see in Figure A.3, you can also browse through user comments using SciTE jump. Another tool that helps navigation through functions is the List Functions tool (Atl + L), which works very similarly to SciTE jump.

The Find tool (Search Menu, Find, or Ctrl + F, Figure A.4) does the common job of finding the desired pieces of text as in most text editors. However, it also enables the marking of the lines containing these pieces of text (click on "Mark all," the result is in Figure A.5), which can make it much easier to study codes made by other people. If you regret what you did, and want to remove the marks, just press Ctrl + R, and they will be gone.

In the search menu, you also find Replace, which has a very similar interface to that of Find (Figure A.6). It is very convenient and allows the replacement of all instances of a given expression in the code. It can be accessed using Ctrl + H.

A.3 Tools that Facilitate Navigation

[Find tool window dialog: Find what: WinActivate; checkboxes for Match whole word only, Case sensitive, Regular expression, Wrap around, Transform backslash expressions; Direction: Up/Down; buttons Find Next, Mark All, Cancel]

Figure A.4 Find tool window.

```
1    Func FAKASGoTo($input)
2        WinActivate("FAKAS")
3        Mouseclick("left",80,360)  ;Input area
4        Send("{BACKSPACE 50}","{DEL 50}")
5        Send($input)
6        Sleep(500)
7        WinActivate("FAKAS")
8        Mouseclick("left",80,380)  ;Go to sample button
9        Sleep(6000)
10   EndFunc
11
12   Func FAKASNeedleDown()
13       WinActivate("FAKAS")
14       Mouseclick("left",80,410)  ;Push Needle Down button
15       Sleep(5000)
16   EndFunc
17
18   Func FAKASNeddleUp()
19       WinActivate("FAKAS")
20       Mouseclick("left",80,440)  ;Pull Needle Up button
21       Sleep(5000)
22   EndFunc
```

Figure A.5 Bookmarks at all WinActivate instances in a piece of code.

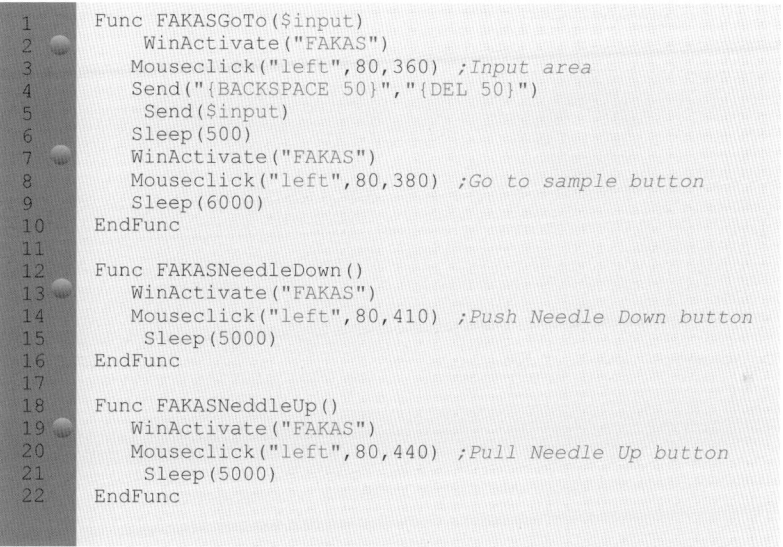

Figure A.6 Replace tool window.

Appendix B

Optical Character Recognition

In Chapter 4, we learned how pixel monitoring can be useful for automation. At the end of the chapter, I briefly mentioned that a more sophisticated pixel analysis was possible, one that allows, among other things, the quantification of a value observed on the screen. For example, if the screen shows the image of "34," instead of getting a value for the pixels forming the image, it is possible to get the number "34." This could be useful, for example, if the background signal of a given instrument is shown on the screen, but is not available for as a value that can be read in a control, or copied to the clipboard.

The technique that enables such kind of character recognition is called optical character recognition (OCR). OCR is widely used when scanning documents. Using OCR, a text document, when scanned, can have the words and numbers in it recognized and made available for searching and copying.

OCR has many important applications, and has been continuously developed since the 1970s. Still, it is an imperfect technology as the task that it tackles can be quite difficult. In particular, recognizing characters on a computer screen is not simple, especially if the characters being shown are small. Therefore, unfortunately nowadays there is no simple OCR solution for this application. Thus, I decided that this topic would not be part of the main portion of the text, because it may not work in many applications. Still, it is possible that a better OCR technology comes in future, and then it is good to learn the basics shown here. Finally, although unlikely, it is possible that the approach presented here works for your specific case, which then would be very fortunate.

B.1 OCR in AutoIt

There are two UDFs for OCR in Autoit. One uses modi, which is a component of Microsoft Office, and the other one uses Tesseract, a project developed by Google. Both are imperfect, and thus I personally cannot see an advantage of one above the other. Since their performance is similar, I will only present the one based on modi, which was developed by Ptrex.

The modi-based OCR needs that either you have Microsoft Office 2003 or 2007 installed on your computer or you download an extra module if you have versions 2010 or 2013. If you do not have any of these, you cannot use this technique. If you have versions 2003 or 2007 installed, you can skip the next paragraphs until the code.

Appendix B Optical Character Recognition

Figure B.1 Shared Point Designer 2007 installation options showing all options marked as Not Available.

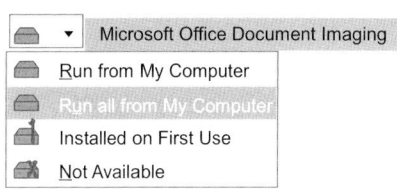

Figure B.2 Shared Point Designer 2007 installation options showing Office Tools, Microsoft Office Document Imaging, and the option Run all from my computer.

If you have versions 2010 or 2013, you need to go to http://www.microsoft.com/en-au/download/details.aspx?id=21581 and download SharePoint Designer 2007. After that, you need to follow these steps in order that the installation is done properly.

First, when in installation options, you need to mark all options as "Not Available" (Figure B.1).

After this, you need to expand "Office Tools" and select "Microsoft Office Document Imaging," there choosing "Run all from my computer" (Figure B.2).

With these options selected, you can proceed with the installation. Once it is finished, we can write our script. The code shown here (Code sample B.1) is a simplification of the one presented by Ptrex in https://www.autoitscript.com/forum/topic/50608-real-ocr-in-au3-in-a-few-lines/?page=1.

```
$path = "figure.tif"
$miDoc = ObjCreate("MODI.Document")
$miDoc.Create($path)
$miDoc.Ocr(9, True, False)
$i = 0
For $oWord in $miDoc.Images(0).Layout.Words
   $palavra = $oWord.text
   ConsoleWrite($palavra& @CRLF)
   $i += 1
Next
```

Code sample B.1 Simple script that gets the characters of a figure using modi OCR.

Before running Code sample B.1, make sure there is a figure file named "figure.tif" in the same folder of the script, or else any other figure file, as long as you assign the correct path to "$path." Also, try to use a figure that has words in it. If you do it right, you will see the words being listed in the console.

I am not going to analyze the code in Code sample B.1 as it involves advanced aspects of AutoIt that are not relevant to the other content explored in this book. In short, it deals with COM objects, which are entities that enable programs to interact with each other. Interested readers can get a primer from https://www.autoitscript.com/wiki/CommAPI.

B.2 Copying from the Screen and Applying OCR

At the start of this appendix, I mentioned that a possible use of OCR would be to get the value of a number presented on the screen. The code below shows how this could be done for FACACO:

```
#include <ScreenCapture.au3>
$path = "screenshot.tif"
opt("WinTitleMatchMode",1)
WinMove("FAke","",0,0)
WinActivate("FAke")
_ScreenCapture_Capture($path, 315, 70, 420, 95)
$miDoc = ObjCreate("MODI.Document")
$miDoc.Create($path)
$miDoc.Ocr(9, True, False)
$i = 0
For $oWord in $miDoc.Images(0).Layout.Words
   $palavra = $oWord.text
   ConsoleWrite($palavra& @CRLF)
     $i += 1
Next
```

Code sample B.2 Copying an area of the screen, in this case, a small field in FACACO interface, and analyzing it using OCR.

Before running the script, make sure you type a number at the topmost results field in FACACO. Running the script, this area of the screen is captured using the function "_ScreenCapture_capture," and is saved to the file in the $path variable (screenshot.tif). Then, the file is analyzed in the same way as the one in Code sample B.1. However, you probably did not get any value using this approach. This is because the number in the figure being analyzed is too small (you can open the file and see that). OCR simply cannot correctly analyze it.

As it is clear by the example provided here, OCR is still not a satisfactory solution for obtaining values from the computer screen. Perhaps future technologies will overcome these limitations.

Appendix C

Scripting with Nonstandard Controls (UIA)

In Chapter 5, you learned how to automate programs using controls, which was very convenient because that way you did not need to use mouse clicks or keyboard entries. Also, the windows being automated did not need to be in the foreground, and thus the automated programs could be used at the same time that other tasks were being done on the computer.

A problem with controls, however, is that they are not always accessible. Many programs, upon examination by AWI (AutoIt Windows Info, more details in Chapter 5), will show nothing that can be used in a script. The reason is that AWI can see controls built using the Windows API as a framework. APIs means Application Programming Interfaces, which is a set of protocols and tools for building software. Many developers, when building their programs for Windows, use the Windows API. However, some do not, and AWI cannot find the controls that would enable automation in these cases.

The alternative approach to enable control-based automation is the use of Microsoft UIA, which stands for user interface automation. UIA is an API, and it is more general in scope than the Windows API. Therefore, more programs have UIA elements than Windows API elements.

Toward the end of 2013, a member of the AutoIt community (junkew), published a code package for AutoIt based on UIA, which enables the access to controls of applications that were built using UIA. This package has tremendous potential for laboratory automation because a number of popular programs controlling laboratory instruments are not built using Windows API (e.g., any program written in LabView, from National Instruments, does not use Windows API, but uses UIA). However, the code package is still under development and may still change considerably, which implies that details of the techniques presented here may change in future for newer versions of the code. This was the reason why I decided to keep this topic out of the main text, which deals with "standard" AutoIt techniques. Nevertheless, readers dealing with software built not using the Windows API can benefit greatly from learning the techniques presented in this appendix. It is assumed that the reader has some knowledge of AutoIt, so it is suggested that you have studied through the book before studying this appendix.

C.1 Downloading the UIA Software Package

The first step to start using UIA in AutoIt is to download the code developed by junkew from https://www.autoitscript.com/forum/topic/153520-iuiautomation-ms-

Figure C.1 Part of SimpleSpy interface, the equivalent of AWI for UIA.

framework-automate-chrome-ff-ie/. Make sure you download at least the latest version of the UIA_VXX.zip (UIA_V0.51 at the last time I downloaded). Decompress the downloaded file, and you will find SimpleSpy.au3 and UIAWrappers.au3 in the file package. These are the files that we are going to use most in the package. Also, if you use Windows XP or Vista, the package may not work unless you also download another package from Microsoft, from the link http://support.microsoft.com/kb/971513.

Before proceeding, it is useful to make a very small modification in the "UIAWrappers.au3" file. Open the file on SciTE, and, on line 560, you should find a line saying "_UIA_LogFile(@YEAR & @MON & @MDAY & "-" & @HOUR & @MIN & @SEC & @MSEC & ".XML", true)." Transform this line in a comment by placing a ";" at its start. This will stop the generation of logfiles for every time you run any script including the functions in this library. This line was kept in the original function because it is a project still under development, and thus it is important to log the execution in order to identify problems. However, this will not be useful for us, and thus it is better to disable it.

Once this modification is done, we can start using the package. The main tool is "SimpleSpy.au3," which is the equivalent of AWI for UIA. SimpleSpy's interface is less polished than that of AWI (Figure C.1).

In order to use SimpleSpy, simply place the mouse pointer on a certain element of interest and press Ctrl + w. For example, open FAKe AutoSampler (FAKAS), place the mouse on the "Go to sample" button, and press Ctrl + w. You should see SimpleSpy listing the relevant information (Figure C.2). Note that you do not need to have SimpleSpy on the foreground of the screen in order that Ctrl + w lists the information of a selected element. This is because Ctrl + w was set as a hotkey, and not an accelerator (see more about the difference in Section 18.2).

SimpleSpy is not a text editor, and thus it is difficult to work with the generated information directly on it. Also, most of the times you need to deal with more than a single control, and SimpleSpy only shows information for a control at a time. Thus, a good strategy is to copy the information provided and paste it into a text editor, like Notepad or even SciTE. After that, you may wish to save the information as an individual file for each control so that it becomes easy to be accessed always that needed. You may also wish to compile SimpleSpy to an .exe file by sending Ctrl + F7 to SciTE, so that you can have it as a stand-alone application while you write other scripts.

C.2 Sending Instructions

As we learned in Chapter 5, one of the two key features of scripting using controls is the ability to send instructions without relying on mouse clicks or keyboard inputs (the other is the ability to get information from controls, which will be examined in the next

```
Mouse position is retrieved 456-238
At least we have an element [Go to sample][Button]

Having the following values for all properties:
Title is: <Go to sample>    Class := <Button>  controltype:= <UIA_ButtonControlTypeId>    ,<
*** Parent Information top down ***
1: Title is: <Desktop>      Class := <#32769>controltype:= <UIA_PaneControlTypeId>, <50033>,    (
"Title:=Desktop;controltype:=UIA_PaneControlTypeId;class:=#32769"
0: Title is: <FAKAS> Class :=<AutoIt v3 GUI>    controltype:= <UIA_WindowControlTypeID>
"Title:=FAKAS;controltype:=UIA_WindowControlTypeId;class:=AutoIt v# GUI"


:~ *** Standard code ***
#include "UIAWrappers.au3"
AutoItSetOption("MustDeclareVars:, 1)

Local $oP0=_UIA_getObjectByFindAll($UIA_oDesktop, "Title:=FAKAS;controltype:=UIA_WindowContr
_UIA_Action($oP0,"setfocus")
_UIA_setVar("Gotosample.mainwindow","title:=Go to sample;classname:=Button")
_UIA_Action("Gotosample.mainwindow","setfocus")


*** Detailed properties of the highlighted element ***
UIA_title:= <Go to sample>
UIA_text:= <Go to sample>
UIA_regexptitle:= <Button>
UIA_class:= <Button>
UIA_regexpclass:= <Button>
UIA_iaccessiblechildid:= <0>
UIA_id:= <7>
UIA_handle:= <199018>
UIA_RuntimeId:= <42;199018>
UIA_BoundingRectangle:= <402;222;100;25>
UIA_processId:= <5556>
UIA_ControlType:= <50000>
UIA_LocalizedControlType:= <button>
UIA_Name:= <Go to sample>
UIA_AccessKey:= <Alt+g>
UIA_HasKeyboardFocus:= <False>
UIA_IsKeyboardFocusable:= <True>
UIA_IsEnabled:= <True>
UIA_AutomationId:= <7>
```

Figure C.2 SimpleSpy showing information about the "Go to sample" button on FAKAS.

subsection). Let us examine how to send mouse clicks and keyboard inputs using the UIA code package.

C.2.1 Mouse Clicks

The first step to send an instruction to a control is to identify its relevant UIA elements, which we learned we can do with SimpleSpy (Figure C.2). The information provided by SimpleSpy is comprehensive, containing much more than what we need to set up a mouse click. It is thus necessary to identify the specific information for our objectives. When this appendix was being written, there was no help file for the UIA package, and thus it was not simple to isolate such specific information. I am not sure if such a help file or tutorial will be available at the time when you read this, so I will here present the easiest ways that I could find to perform what we need, which in the present case is a mouse click.

In the information obtained by SimpleSpy for the Go to sample button (Figure C.2), you can find a section named "Standard code" (Code sample C.1). This is a mini-script that you can copy and paste to SciTE and test whether you can reach the desired control.

```
1) #include "UIAWrappers.au3"
2) AutoItSetOption("MustDeclareVars", 1)
3) Local $oP0=_UIA_getObjectByFindAll($UIA_oDesktop,
"Title:=FAKAS;controltype:=UIA_WindowControlTypeId;class:=AutoIt v3
GUI", $treescope_children)
4) _UIA_Action($oP0,"setfocus")
5) _UIA_setVar("Gotosample.mainwindow","title:=Go to
sample;classname:=Button")
6) _UIA_action("Gotosample.mainwindow","setfocus")
```

Code sample C.1 Standard code generated by SimpleSpy for the Go to sample button on FAKAS.

You can try running the script (remove line numbering first), and should see FAKAS coming to the foreground, and being involved by a red rectangle. However, the button will not be involved by a rectangle, meaning that it was not recognized. The red rectangle is a standard characteristic of the UIA package, being a useful feature that helps us know whether the script is really selecting the desired elements. Unfortunately, as the lack of the Go to sample button illustrates, the standard code generated by SimpleSpy is not fully functional in some cases, and we need to tweak it. Before doing that, let us have a look on it. First, it includes the UIAwrappers.au3 file, which contains the functions used in the code. As you know, in order to access the code, you will need to have this file in the same folder of your scripts or a system folder, as shown in Section 10.1. The next line is not necessary, it determines that all variables must be declared. This may be good programming practice for long code, but, as we know, in this book scripts are quite short. So, I am going to remove this line for the next examples.

The following line in Code sample C.1 introduces a function of UIAwrappers.au3 library: "_UIA_getObjectByFindAll." The first argument passed to the function is "$UIA_oDesktop," which is a constant that refers to the desktop. The desktop contains all windows being shown, and is the biggest "window" on the computer in this sense. The next argument is a long string: "Title:=FAKAS;controltype:=UIA_WindowControlTypeId;class:=AutoIt v3 GUI." This is the key information that tells the function which window we want to automate. Sometimes, you can omit part of this information, and provide only the title or only the class, for example. However, providing the full information is the best way because it works in all cases, and thus we are going to keep adopting this approach through the rest of the text. The third parameter is the constant "$treescope_children," which determines that the scope of the search for the function are the children windows of the desktop. Finally, the function returns a value, which is passed to the variable "$oP0." This means that for other functions, $oP0 can be passed as the reference for FAKAS windows. _UIA_getObjectByFindAll is one of the most important functions in the UIA package, and we will use it in all the following examples.

The next function in the code is "_UIA_action," which takes as arguments the FAKAS windows, by means of the variable $oP0, and an instruction, "setFocus." This instruction tells the window to be in the foreground. This is necessary in many cases, and will be adopted as a standard practice. In other words, it is not possible to have everything working in the background when dealing with UIA. _UIA_action is arguably the most important function when using UIA to automate programs. We will deal with _UIA_action all the time in this appendix.

The next line contains the function "_UIA_setVar." This function creates a variable that can be passed to the next function, _UIA_action. It works a little differently from the other variables in AutoIt, because the created variable does not use the "$," that is used for all other variables. The first argument is not really an argument, but the name of the variable being created. For some unclear reason, the proposed name, "Gotosample.mainwindow" does not work. It is possible to change this name in both this and the next function to make the code work. The next parameter is the identifier of the control that we want to access, the Go to sample button. In this case, the identifier was the title of the control (Go to sample). Finally, the type of control, identified by "classname," is passed. In this case, we have a "Button." Because of its strange syntax, I am not particularly fond of _UIA_setVar, and prefer using a different function to assign a variable to a control in a window. I will get to it at the appropriate time.

Finally, the last line contains again _UIA_action, but this time for "Gotosample.mainwindow." This does not work, as we already know. However, as explained in the previous paragraph, if we substitute this parameter with a proper value, it will work. Let us see the working code below:

```
1) #include "UIAWrappers.au3"
2) $oP0=_UIA_getObjectByFindAll($UIA_oDesktop,
"Title:=FAKAS;controltype:=UIA_WindowControlTypeId;class:=AutoIt v3
GUI", $treescope_children)
3) _UIA_Action($oP0,"setfocus")
4) _UIA_setVar("GoTo","title:=Go to sample;classname:=Button")
5) _UIA_action("GoTo","invoke")
```

Code sample C.2 Modified and improved code based on the standard code generated by SimpleSpy for the Go to sample button on FAKAS.

Remove line numbering, run the code, and you should see the Go to sample button being activated this time. The differences between Code samples C.1 and C.2 are that the unnecessary lines (as commented in the previous paragraphs) were removed, and the last instance of _UIA_action now receives as arguments "Go to sample," which is the button label, and "invoke," which is the instruction that results in a mouse click. You may want to know that instead of invoke, we could have used "click" and the code would have worked in the same way. However, click is less powerful than invoke because it makes actual use of the mouse pointer. Thus, if the mouse is being used for another purpose, this could interfere with the proper execution of the script. Therefore, invoke is preferable to click.

Although the example in Code sample C.2 worked, I am in favor of a different approach to using UIA. As I mentioned before, I am not comfortable with the syntax used to connect _UIA_setVar and _UIA_action. I think it is preferable to use a function that defines a real AutoIt variable (marked with a "$") for a control in the window. Fortunately, such a function exists, and, even better, the variable created using it can be passed to _UIA_action as an argument. Let us see the code presenting it:

```
1) #include "UIAWrappersMinha.au3"
2) $FakasWindow=_UIA_getObjectByFindAll($UIA_oDesktop,
"Title:=FAKAS;controltype:=UIA_WindowControlTypeId;class:=AutoIt v3
GUI", $treescope_children)
3) $GoToSampleButton = _UIA_getFirstObjectOfElement($FakasWindow,
"title:=Go to sample", $treescope_subtree)
4) _UIA_action($GoToSampleButton,"invoke")
```

Code sample C.3 Introducing _UIAgetFirstObjectOfElement.

If you run the script (again, remove line numbers), you should see it performing like the one in Code sample C.2. Note that the code lines are long and there are only four lines here: the first with the include, a long one defining the "$FakasWindow" variable, the next one defining the "$GoToSampleButton" variable, and the last one calling _UIA_action. This time, instead of using _UIA_setVar, we used "_UIAgetFirstObjectOfElement." This function receives as its first argument the window containing the element, which was defined in the previous line _UIA_getObjectByFindAll. The next argument is the identifier of the control, which was its title. The final argument is the constant "$treescope_subtree," which sets the scope of the search in the function for the subtree inside the window. This constant, and also $treescope_children, which was passed to _UIA_getObjectByFindAll, are defined in the file CUIAutomation.au3, and should not be an object of concern for the simple scripts presented here. Simply include them as parameters for these functions, and they should work correctly. Finally, the function _UIA_action is called, this time receiving the variable created by _UIA_getFirstObjectOfElement as an argument. Code sample UIA3 will be the template for all the following codes in this appendix.

C.2.2 Keyboard Inputs

Let us now see how to send keyboard inputs. Let us select an input field, like the sample number input area in FAKAS (Figure 1.5). Again, the first step is to use SimpleSpy and gather the relevant information. Doing so, and looking at the Standard code, we find that the title for the input field is "Needle is up":

```
1) #include "UIAWrappers.au3"
2) AutoItSetOption("MustDeclareVars", 1)
3) Local $oP0=_UIA_getObjectByFindAll($UIA_oDesktop,
"Title:=FAKAS;controltype:=UIA_WindowControlTypeId;class:=AutoIt v3
GUI", $treescope_children)
4) _UIA_Action($oP0,"setfocus")
5) _UIA_setVar("Needleisup.mainwindow","title:=Needle is
up;classname:=Edit")
6) _UIA_action("Needleisup.mainwindow","setfocus")
```

Code sample C.4 Standard code generated by SimpleSpy for the input area on FAKAS.

This looks strange, and in fact it is. For example, press the Push needle down button, and run SimpleSpy again. Now the title of the control will be "Needle is down." Apparently, the input area is taking the value of the label at the bottom of FAKAS, which tells the status of the "autosampler." The cause is unclear, and will not be explored here. Anyway, for practical purposes, this change in the control title depending on the status of the autosampler makes it complicated to use this parameter in a script. Fortunately, it is simple to solve this problem. Besides the standard code, SimpleSpy provides a wealthy of information about each control (Figure C.2). A very important one is the "AutomationId," which you can find among the elements in the list named "*** Detailed properties of the highlighted element ***." For the input area, the AutomationId is "6." Knowing this, we can rewrite Code sample C.3 (Code sample C.4 will not be used anymore):

```
1) #include "UIAWrappers.au3"
2) $FakasWindow=_UIA_getObjectByFindAll($UIA_oDesktop,
"Title:=FAKAS;controltype:=UIA_WindowControlTypeId;class:=AutoIt v3
GUI", $treescope_children)
3) $FakasInput = _UIA_getFirstObjectOfElement($FakasWindow,
"AutomationId:=6", $treescope_subtree)
4) _UIA_action($FakasInput,"sendkeys","{DEL 50}")
5) _UIA_action($FakasInput,"sendkeys","{BACKSPACE 50}")
6) _UIA_action($FakasInput,"sendkeys","4")
```

Code sample C.5 Sending keystrokes to the input area in FAKAS.

Run the script, and you should see the input area of FAKAS being erased and replaced with "4." This code is very similar to Code sample C.3, the main differences being that the second argument for _UIA_getFirstObjectOfElement was the automation ID, and not the title, and that the invoke instruction was replaced by "sendkeys" when calling _UIA_action, and there was a third argument for this function as well, which consisted in the contents of the sendkeys instruction. If you have been following the book, you should be familiar with the Send function, and should have realized that sendkeys and Send work in the same way.

You could be thinking that it is best to always rely on the automation ID instead of the title of a control. This is not true. Sometimes, the automation id is not available, and you need to use the title. There is not a rule of thumb, only trial and error can determine the best approach to be adopted in each case.

C.3 Getting Information about Controls

The other necessary task in order to enable full automation using controls is to get information about them, so that interactive scripting becomes possible (and does not rely on pixel monitoring). I found, however, that obtaining information from controls was not trivial using the UIA package when writing this book. So, I modified the original UIAWrappers.au3 file by adding the following lines of code:

```
1) case "PickName"
2)    $retVal=_UIA_getPropertyValue($obj2ActOn, $UIA_NamePropertyId)
3) case "PickValue"
4)    $retVal=_UIA_getPropertyValue($obj2ActOn,
$UIA_ValueValuePropertyId)
5) case "IsEnabled"
6)    $retVal=_UIA_getPropertyValue($obj2ActOn,
$UIA_IsEnabledPropertyId)
7) case "IsSelected"
8)    $retVal=_UIA_getPropertyValue($obj2ActOn,
$UIA_SelectionItemIsSelectedPropertyId)
```

Code sample C.6 Lines of code to be added to UIAWrappers.au3 in order to enable easier retrieving of information from controls.

Open the UIAWrappers.au3 file and go to line 1440, which should read "_UIA_ DumpThemAll($obj2ActOn,$treescope_subtree)." This is one of the possible actions taken by _UIA_action as determined by a long Switch conditional (for more details on Switch conditionals, see Section 18.2). Go to the end of this line and press ENTER, so that you start typing on line 1441. There you should add Code sample C.6 (without line numbers), and save the file with another name, which here I chose to be "UIAwrappersNew.au3." By adding these lines of code, you enable easier retrieving of information from some controls than what is enabled by default in the UIA package, as the following examples will show.

Before getting to the examples, let us have a quick look at the key element of the added lines of code, which is the _UIA_getPropertyValue function. This function receives two variables as arguments. The first is "$obj2ActOn," which refers to the element that is passed as argument for _UIA_action (e.g., $FakasInput in Code sample C.5). The other one can be anyone among the many listed in the "$UIA_propertiesSupportedArray," which is defined in the UIAWrappers.au3 file on line 137. Here, only four of them are presented: "$UIA_NamePropertyId," "$UIA_ValueValuePropertyId," "$UIA_IsEnabledPropertyId," and "$UIA_SelectionItemIsSelectedPropertyId." Therefore, if one of the words listed as possible inputs for "case" is passed as an argument for the function _UIA_action, the corresponding value is passed to the variable "$retVal." For example, if "Pickname" is used as an argument for _UIA_action, $retVal receives a value corresponding to the variable $UIA_NamePropertyId. It may be easier to understand by examining an example, which will be presented in the next subsection for FAKAS.

C.3.1 Getting Information from FAKAS Controls

Let us begin with FAKAS, since it is a simple and familiar interface:

```
1) #include "UIAWrappersNew.au3"
2) $FakasWindow=_UIA_getObjectByFindAll($UIA_oDesktop,
"Title:=FAKAS;controltype:=UIA_WindowControlTypeId;class:=AutoIt v3
GUI", $treescope_children)
3) $GoToSampleButton = _UIA_getFirstObjectOfElement($FakasWindow,
"automationid:=7", $treescope_subtree)
4) $ButtonLabel = _UIA_action($GoToSampleButton,"PickName")
5) ConsoleWrite($ButtonLabel&@CRLF)
```

Code sample C.7 Retrieving the label from the Go to sample button in FAKAS.

Run the script, and you should see some action being done on FAKAS (as the red rectangles make us aware), and also the text "Go to sample" appearing in the console of SciTE. Code sample C.7 should be familiar by now: we define the window to be automated using _UIA_getObjectByFindAll, then the Go to sample button using _UIA_getFirstObjectOfElement. Note that the automation ID, and not the title, is used to identify the button. After that, _UIA_action is called using $GoToSampleButton and "PickName" as arguments. As we learned when analyzing Code sample C.6, by passing PickName as an argument to _UIA_action we determine that $UIA_NamePropertyId is returned by _UIA_action. Therefore, in our code, the variable "$ButtonLabel" receives this value, which is "Go to sample" for our case, and is printed in the console area of SciTE by means of the function ConsoleWrite.

Let us now see another example:

```
1) #include "UIAWrappersNew.au3"
2) $FakasWindow=_UIA_getObjectByFindAll($UIA_oDesktop,
"Title:=FAKAS;controltype:=UIA_WindowControlTypeId;class:=AutoIt v3
GUI", $treescope_children)
3) $FakasInput = _UIA_getFirstObjectOfElement($FakasWindow,
"automationid:=6", $treescope_subtree)
4) For $i = 1 to 10
5)    _UIA_action($FakasInput,"sendkeys","{DEL 50}")
6)    _UIA_action($FakasInput,"sendkeys","{BACKSPACE 50}")
7)    _UIA_action($FakasInput,"sendkeys",$i)
8)    ConsoleWrite(_UIA_action($FakasInput,"PickValue")&@CRLF)
9) Next
```

Code sample C.8 Retrieving the value in the input area in FAKAS.

Run the script, and you should see the value in the input area in FAKAS changing from 1 to 10, and the same values being printed in the Console of SciTE. As you can see, Code sample C.8 is a combination of Code samples C.5 and C.7. By passing "PickValue" as an argument to _UIA_action, and by then using its return value directly as an argument for ConsoleWrite, we could retrieve the values inside the input area and print them in the console.

One more example using FAKAS:

```
1) #include "UIAWrappersNew.au3"
2) $FakasWindow=_UIA_getObjectByFindAll($UIA_oDesktop,
"Title:=FAKAS;controltype:=UIA_WindowControlTypeId;class:=AutoIt v3
GUI", $treescope_children)
3) $GoToSampleButton = _UIA_getFirstObjectOfElement($FakasWindow,
"automationid:=7", $treescope_subtree)
3) ConsoleWrite(_UIA_action($GoToSampleButton,"IsEnabled")&@CRLF)
_UIA_action($GoToSampleButton,"invoke")
4) ConsoleWrite(_UIA_action($GoToSampleButton,"IsEnabled")&@CRLF)
```

Code sample C.9 Retrieving the Go to sample button status in the input area in FAKAS.

Running the script, the result is very similar to that using Code sample C.8, but the values appearing in the console are "True" and "False" this time. This time, the argument for _UIA_action is "IsEnabled," which in this case returns True if the button is enabled, and False if it is not enabled. The button was enabled at the start of the script because it was not clicked. However, after being clicked (by passing invoke to _UIA_action), the button becomes unavailable for clicking, and thus its status regarding IsEnabled becomes False.

Yet another example for FAKAS:

```
1) #include "UIAWrappersNew.au3"
2) $FakasWindow=_UIA_getObjectByFindAll($UIA_oDesktop,
"Title:=FAKAS;controltype:=UIA_WindowControlTypeId;class:=AutoIt v3
GUI", $treescope_children)
3) $GoToSampleButton = _UIA_getFirstObjectOfElement($FakasWindow,
"automationid:=7", $treescope_subtree)
4) $FakasStatus = _UIA_getFirstObjectOfElement($FakasWindow,
"automationid:=5", $treescope_subtree)
5) _UIA_action($GoToSampleButton,"invoke")
6) For $i = 1 to 15
7)    ConsoleWrite(_UIA_action($FakasStatus,"PickName")&@CRLF)
8)    Sleep(500)
9) Next
```

Code sample C.10 Retrieving the status field value in FAKAS.

Run the script, and this time you should see the console displaying: "Needle is up, Moving to sample (nine times), Arrived at sample (two times), Needle is up (three times)." As you can see in the code, a loop at the end retrieves at every 0.5 s the contents of the status field at the bottom of the FAKAS interface, which changes reflecting what the program is doing.

C.3.2 Getting Information from Controls of Other Programs

Now, let us see an example for a different program, one that has radio buttons. We can use Code sample 5.9, for example. Run it as an executable (by compiling it using Ctrl + F7 on SciTE), and launch SimpleSpy to get the relevant arguments for _UIA_getObjectByFindAll and _UIA_getFirstObjectOfElement. Put the mouse on the radio button marked as "Up" and press Ctrl + W, so that SimpleSpy gets the information we are going to use here. Doing so, we can use the following code to get the status of a radio button:

```
1) #include "UIAWrappersNew.au3"
2) $CrioWindow=_UIA_getObjectByFindAll($UIA_oDesktop,
"Title:=Crio;controltype:=UIA_WindowControlTypeId;class:=AutoIt v3
GUI", $treescope_children)
3) $RadioUp = _UIA_getFirstObjectOfElement($CrioWindow,
"automationid:=4", $treescope_subtree)
4) ConsoleWrite(_UIA_action($RadioUp,"IsSelected")&@CRLF)
5) _UIA_action($RadioUp,"invoke")
6) ConsoleWrite(_UIA_action($RadioUp,"IsSelected")&@CRLF)
```

Code sample C.11 Retrieving the status of a Radio Button.

Run the script. If the radio button was first blank, you should see "False" displayed in the console. Alternatively, if it was black at the start, you should see "True" in the console. The argument "IsSelected," passed to _UIA_action, makes this function return the value corresponding to the status of the radio button, which can be False, if it is blank, or True, if black.

By this point, you could be thinking that each different type of control may have a different variable associated with it, and that this variable can be passed to _UIA_action in order to return a proper attribute determining the status of this control. This is true. For example, there are variables associated with items in a Combo box or menu items. A list of such items would be exhaustive, and thus I think it is better to show how you can discover which item you need to use instead of providing such a list. Let us see menu items, for example. Run Notepad and go to the "Edit" menu. The first option, "Undo," should be inaccessible, since you just started the application. Go to SimpleSpy, get the information about Undo, and paste it in SciTE, for example. Then, on Notepad, again type anything and then open the Edit menu, now with the Undo option available. Use SimpleSpy again to get its data, copying them to SciTE, but not on top of the other data about the unavailable Undo. By doing this, we can now compare the same control in two different states. Because SciTE lines are numbered, it is easy to search for small differences between two very similar pieces of text. In this case, we find that at line 45 the two texts differ: for the unavailable Undo, the text reads "UIA_IsEnabled:= <False>," while for the available one it reads "UIA_IsEnabled:= <True>." Thus, it is clear that

Table C.1 Properties returned by SimpleSpy and their corresponding variables and instructions to be passed to _UIA_action in UIAwrappersNew.au3.

SimpleSpy property	UIA wrappers.au3 variable	Instruction for _UIA_action
UIA_Name	$UIA_NamePropertyId	PickName
UIA_ValueValue	$UIA_ValueValuePropertyId	PickValue
UIA_IsEnabled	$UIA_IsEnabledPropertyId	IsEnabled
UIA_SelectionItemIsSelected	$UIA_SelectionItemIsSelectedPropertyId	IsSelected

Restricted to cases used in Code sample C.6.

it is possible to differentiate the states of this menu item by using the UIA_IsEnabled property, which in Code sample C.6 corresponds to $UIA_IsEnabledPropertyId, and thus to the instruction IsEnabled that can be passed to the function _UIA_action. See Table C.1 for the corresponding properties returned by SimpleSpy, their corresponding variables in UIAwrappers.au3, and the instructions to be passed to _UIA_action in our new modified version, UIAwrappersNew.au3.

It is possible that you find a case that is not covered by the options in Code sample C.6. In this case, you need to write the code yourself. However, you can see the pattern is very clear: a SimpleSpy property corresponds to a UIAwrappersNew.au3 variable that has the "$" at the start of its name, the name of the property in SimpleSpy in the middle, and "PropertyId" at its end. The instruction to be passed to _UIA_action you must define yourself, following the structure presented in Code sample C.6.

C.4 Automating a LabView Program

FAKAS can be accessed using the standard control functions in AutoIt (see Chapter 5), so it is not the best example to demonstrate how UIA can be useful. Let us see a case in which UIA can access controls, but AWI (AutoIt Window Info) cannot. Typical cases are programs built using the LabView language, which are very common in some laboratory environments. If there is a program written in LabView installed on a computer in your laboratory, you can test with it. I have found cases in which even UIA will not work. So, let us see a case in which it works, so that the point can be made fully clear, and all readers can test the UIA package.

Download the free program available at http://www.thorlabs.hk/software_pages/viewsoftwarepage.cfm?code=LCC25, which works for the LCC25 liquid crystal controller, from Thorlabs. Install the program and launch it. Then, you can test UIA and AWI in order to get information on the controls of the program. Using AWI, you will not be able to obtain much information about the program, except the window title, and a few more data about the window as a whole. By contrast, using SimpleSpy, you will be able to obtain useful information about all controls on the interface. For example, place the mouse pointer over "connect," at the top left corner (Figure C.3), and press Ctrl + w. You will obtain, among other things, the automation ID for this control, which you learned that can be used to send a mouse click, for example. Try on other elements, and you should find some suitable values for reading too.

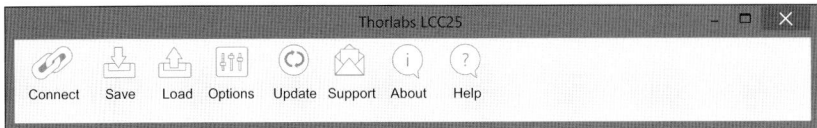

Figure C.3 Tool bar on Thorlabs LCC25 interface.

C.5 Summary

- There is a software package for AutoIt based on UIA that allows the access to control information.
- In some cases, UIA gives more information than the standard AutoIt tool, the AWI.
- Despite being powerful and useful, the UIA package is still being developed and can change considerably.
- Another aspect of the novelty of the UIA package is that there is no help file available for it (at the time of writing this book at least).
- SimpleSpy is the equivalent of AWI for UIA, giving the necessary information about the controls to be automated.
- _UIA_getObjectByFindAll is a function that identifies the window to be automated.
- _UIAgetFirstObjectOfElement is a function that identifies a control inside a window.
- _UIA_action is the most important function in the UIA package, and can be used to send instructions to, and get information from controls.
- Invoke is the best instruction passed to _UIA_action to send a mouse click.
- Sendkey is the instruction passed to _UIA_action to send keyboard inputs.
- A modification was made to the UIA package to facilitate retrieving information from controls.
- Using this modification, PickName, PickValue, IsEnabled, and IsSelected could be passed as instructions to _UIA_action in order to obtain data from controls.
- In order to determine which instruction needs to be used to get the desired information from a control, it is useful to compare a single control in two different states.
- It is possible to further modify the UIA package in order to accommodate customized retrieving of control information.
- Some programs written in LabView have controls that can be accessed using UIA but not the standard tool, AWI.

Index

a

Accelerators 164
Alt 33
Application program interface (API)
    see Windows, API
Arduino 76, 152, 157, 158, 198
Arrays 79
    Ubound 82
ArtSoft 72
AU3Recorder 26, 28, 34
AutoIt
    alternatives 14
    download 15
    history 13
AutoIt v3 Windows Info (AWI) 37, 46, 49, 50, 59, 111
Autosampler 3, 8, 9, 25, 45, 62, 71–73, 75, 76, 198, 216

b

Bingo's chat 144–148
Blocking access 46
BlockInput 46, 47, 63, 69
Boolean variables 68

c

Calc
    library 86
    spreadsheets 92, 154–156
Code, organizing 29, 202
Code wizard 201
COM ports 149, 167, 170
    functions 150
    protocols 149
Comma separated value (.csv) 84

Command prompt 135, 136
Comments 15, 128
Compiling to .exe 58, 109, 130, 163, 212
Computer numeric control (CNC) routers 71
Concatenator 43
Conditionals 55, 128, 130, 163, 165
Console 18
Controls 25, 32, 33, 197, 198
    advantages 54
    definition 49
    getting information 49
    limitations 54
    sending instructions 52
Coordinates on screen 16, 22, 28
CRLF 140, 155, 173
Ctrl 32, 164

d

Data mining 88
Data types 82
Databases 95
    relational 101
Dates 107
DC motors 76
Disk Operating System (DOS) 14
    see also Command prompt
Do it yourself (DIY) devices 77, 198
Do … Until loops 184, 190
Drawing 183, 185, 186, 190

e

Email 61–65, 117
    Simple Mail Transfer Protocol (SMTP) 63

Error check when coding 18
Excel-functions 88, 91
Excel-spreadsheets 83
Excel.au3 84

f
FAKe AutoSampler (FAKAS) 9
FakeCarbonAnalyzerController
    (FACACO) 8, 9
File 87
  listing 88
  reading 120, 139, 140
  sharing 137–139
  writing 139
Flow of action 31, 38, 42,
    45, 55, 123
For ... Next loops 23–25
Functions 16, 29–32

g
G-code 71, 73, 74
General Purpose Interface
    Bus (GPIB) 157
Gmail 62, 64
Graphical user interface
    (GUI) 49, 159
  buttons 161, 162
  combo boxes 169
  COM ports 167–174
  creating 159–161,
    191–194
  editor 191
  graphics 183
  input areas 161, 162
  labels 161, 162
  keyboard shortcuts 163–165
  KODA coding 191, 194, 195
  multitasking 177, 187, 195
  OnEventMode 177, 187, 195
  radio buttons 167, 168
  tabs 173, 174
Graphics device interface plus
    (GDIplus) 183
  animation 185
  functions 183
GuiConstants.au3 189
GuiConstantsEx.au3 160, 189

h
Hotkeys 164

i
If ... Then 55
Image identification 43
Include 32
Indentation 24
Infinite loops 57–59
InputBox 40
Instant messaging 117
    IRC 125
    Trillian 117
Internet Relay Chat (IRC) 125
Instrument integration 2
IP adress 135–137
Ipconfig 135
IrfanView 90

k
Keyboard automation 23, 52, 216

l
LabView 198, 221, 222
Libraries 32
LibreOffice 83
Linux 18
Local area network (LAN) 135
  connecting 135–137, 143–144
  shared folder 137–139
  third party software 143, 148
Loop 23, 40, 87, 94

m
Mach-3 72, 74
Macros 107, 108, 140
Magnifying glass 38, 39
Mathematical Operators 39, 40, 43
mIRC 127
Modular automation 1
Monitoring in real time 87, 153–156,
    185–189
Mouse automation 15–17, 52, 212
MsgBox 24

n
Ncpa.cpl 135
Network protocols 64, 65, 125, 135

o

Open-source hardware 1, 2, 77, 199
Opera web browser 62
Opt 22, 122, 177
Optical character recognition (OCR) 207
Organizing figure files
   color 90
   shape 91

p

Parsing 81
Phone calls 66–69
Pixel
   get color 37, 38
   monitoring 37–40
   screen–area monitoring 43
Pop-up windows *see* InputBox; MsgBox

r

Random numbers 92, 93
Raw data 80
Regular expressions 82
Remote control software 111
Restek ProFlow 6000 flowmeter 155, 188, 189
Robotic arm 76
RS232 protocol 149

s

SciTE 15, 16, 201
   console 18
   download 15
   navigating 203–205
   organizing code 202, 203
   writing aids 17
Scripting 13
Scripting in laboratory automation 3–5
Select … case 130
Semi-automated procedure 90
Serial communication 149–150
Shift 33
Short Message Service (SMS) 65
Sikuli 14
Simple Mail Transfer Protocol (SMTP) 63, 69
Skype 65–68
Sleep 21, 24, 33, 34
SoundPlay 67, 69
Spreadsheet 83
   Calc 83
   Excel 83
   generic 86
SQlite 95
   AutoIt 95, 96
   browser 96, 97, 99, 100, 102, 103
   checking version 95
   creating 96
   download 96
   modification 99
   queries 102
Standards for laboratory automation 3–5, 197, 198
Stopping scripts 25, 26
String 80
String processing 81, 121, 128, 129, 147, 150–153, 173
Structured Query Language (SQL) 97
Switch… case 128, 163, 165

t

TeamViewer 111, 148
3D-printers 71, 75, 198
Time (hour, minute, second, etc) 107, 108
Time macros 107, 108
Timming 21, 24, 33, 34, 67, 68, 151, 152
Total Laboratory Automation (TLA) 1
Trillian 117, 127, 139, 145, 148
Transmission Control Protocol (TCP) 126, 128, 132

u

Universal SERIAL Bus (USB) 149
User defined functions (UDF) 32, 63, 80, 95, 125, 150, 183, 207
User interface automation (UIA)
   definition 211
   download 211
   getting information from controls 217
   keyboard inputs 216
   LabView 221
   modifications 212, 217
   mouse click 213
   SimpleSpy 212

V

Variables   24
Virtual Instrument Software Architecture (VISA)   157
Visual Basic for Applications (VBA)   14

W

What you see is what you get (WYSIWYG)   191
While loops   39, 40

Window
   activation   22, 23
   monitoring   35–37
   positioning   22
   title   22
   waiting   35, 36
Windows API   83, 178, 197, 198, 211